Principles of Qualitative Inorganic Analysis: Precipitation, Separation and Identification of Cations

Authored by

Huda S. Alhasan

Environmental Research and Studies Centre
University of Babylon
Babylon, Hilla
Iraq

&

Nadiyah Alahmadi

Department of Chemistry
University of Jeddah, College of Science
Jeddah
Saudi Arabia

Principles of Qualitative Inorganic Analysis: Precipitation, Separation and Identification of Cations

Authors: Huda S. Alhasan and Nadiyah Alahmadi

ISBN (Online): 978-981-14-9263-1

ISBN (Print): 978-981-14-9262-4

ISBN (Paperback): 978-981-14-9264-8

need for a court order if at any point you breach any terms of this License Agreement. In no event will any delay or failure by Bentham Science Publishers in enforcing your compliance with this License Agreement constitute a waiver of any of its rights.

3. You acknowledge that you have read this License Agreement, and agree to be bound by its terms and conditions. To the extent that any other terms and conditions presented on any website of Bentham Science Publishers conflict with, or are inconsistent with, the terms and conditions set out in this License Agreement, you acknowledge that the terms and conditions set out in this License Agreement shall prevail.

Bentham Science Publishers Pte. Ltd.
80 Robinson Road #02-00
Singapore 068898
Singapore
Email: subscriptions@benthamscience.net

BENTHAM SCIENCE

CONTENTS

FOREWORD

Analytical chemistry is divided into qualitative and quantitative analysis. Qualitative analysis is a sequential set of steps in order to characterize the type of substances that are included in the components of the mixture. The qualitative inorganic analysis is used to demonstrate the containment or non-containment of inorganic elements, ions, or compounds in an unknown sample, for example, an analyst working in the environmental field may test a water sample to confirm whether this sample contains dissolved ions of mercury, lead, or barium.

Nowadays, advanced laboratories and scientific companies perform many qualitative analyses effectively using methods, such as infrared and mass spectroscopy, nuclear magnetic resonance, neutron activation analysis, X-ray diffraction, spectroscopy, chromatography, electrophoresis, and others, despite that qualitative analysis with traditional methods is still very important, whether for field tests or the initial examination of instruments, as well as for students in the early stages of learning. This book is divided into ten chapters. The first chapter describes the method and laboratory safety. The second chapter includes an explanation of the basic principles of qualitative analysis, while chapters 3-8 cover several topics that are important in understanding how a particular analytical method works for diagnosing cations in samples, starting from sedimentation and ending with the identification. As for the ninth chapter, it included a summary of diagnosing cations with some given questions about all chapters. Finally, the book concludes with many important tables and information for both students and researchers.

Although this book is primarily designed for students in the first stage and academic teachers, it may also prove useful in teaching staff and workers in different fields of science and industry, because it contains scientific information that helps in understanding the subject of inorganic cations in analytical chemistry. We hope that the book fulfilled the aspirations of the authors.

Prof. Naser Abdulhasan Almatwari
Department of Chemistry
Faculty of Science / University of Kufa
Iraq

PREFACE

In this practical book, we tried to complete the basics of analytical chemistry. The first chapter describes the laboratory methods and chemical safety with an explanation of how to write the laboratory report. The next five chapters follow the detection of acidic fissures in simple solid compounds or in mixtures. It provides the most important interactions of common cations, which are divided into five groups depending on solubility in a particular solvent. We have taken into consideration the presentation of this book for a student with a detailed explanation of the theoretical basis for each group, which is carried out by the student so that he can recognize the importance of practical experiments and their relationship with theoretical study, while Chapter VII has included observations and general questions. This book is important for undergraduate students (the first class) for each Faculty of Science and some departments of the faculties of engineering and all faculties of the medical group as well as medical institutes. Therefore, we considered writing this book, which is a modest effort to advance the wheel of authorship in the Arab world, in general, and Iraq and Saudi Arabia in particular.

In conclusion, I would like to thank the Ministry of Higher Education and Scientific Research and the University of Babylon/Environmental Research and Studies Centre in Iraq, and the University of Jeddah in Saudi Arabia for providing opportunities to write this book and their continued support for scientific and educational studies.

CONSENT FOR PUBLICATION

Not applicable.

CONFLICT OF INTEREST

The author declares no conflict of interest, financial or otherwise.

ACKNOWLEDGEMENTS

Declared none.

Huda S. Alhasan
Environmental Research and Studies Centre
University of Babylon
Babylon, Hilla
Iraq

&

Nadiyah Alahmadi
Department of Chemistry
University of Jeddah, College of Science
Jeddah
Saudi Arabia

CHAPTER 1

General Guidelines for Laboratory Work

Abstract: Chapter one gives the students the information that they need to bear in mind before entering the laboratory. Moreover, students will learn about the safety rules and be familiarized with the laboratory techniques and apparatus. Chapter one will give instructions about recording observations and writing laboratory reports.

Keywords: Apparatus, Laboratory reports, Safety, Work techniques.

GENERAL GUIDELINES FOR LABORATORY WORK

Instructions for Students

1. The student should wear a lab coat to protect his clothing from pollution and corrosion due to handling concentrated acids and chemicals.
2. Do not smoke and eat food and drink in the laboratory at all.
3. Ensure, upon entering the laboratory, the safety of all electrical and gas connections and the operation of air vacuums.
4. Determine in advance what you want to run from experiments and according to known and specific steps.
5. When the acid is diluted, add the acid to the water in small quantities gradually and make sure that the heat of the diluted solution is not warm (because the process is Exothermic).
6. Use Rubber pipetting bulbs with a pipette when you transfer chemical solutions.
7. When you want to get rid of acid or alkaline solution residues, open the water in the drainage basin before you pour it in and follow with strong water flow.
8. Do not waste the exhaust of chemical compounds that can interact with the water pipes in the drainage basin, but put them in special baskets or containers to be executed in appropriate ways.
9. Be sure to read what is written on the bottles containing chemicals, whether their names or compositions or instructions for how to use or warning of certain risks associated with their use.
10. Return the reagent bottles to their locations after you use them and make sure they are closed.
11. Take only the chemicals you need during weight without loss or waste, and so on for solutions.

12. Take notes about the experiment and write your results.
13. After finishing work, return the equipment that you used and clean your place well and make sure to close the gas socket you used [1, 2].

GUIDELINES FOR WORKING INSIDE THE LABORATORY

1. Use small quantities for reagents and chemicals, may be one drop or two drops of reagent are enough for the interaction to get the desired result.
2. The experiments may be a failure as the result of the contamination of your equipment or the glassware that you use; therefore, make sure you clean your apparatus before starting experiments.
3. All necessary precautions should be taken when using toxic and flammable materials. Be careful not to expose flammable material such as benzene or ether to direct flame. This may lead to a fire.
4. All reactions that result in the rise of vapors or gases that are harmful, toxic or irritating to the lungs or eyes, must be conducted in the fume hood.
5. Place only the necessary equipment and the glassware on the laboratory bench that you use in the laboratory.
6. When heating a solution or liquid in a test tube, do not direct a test tube toward yourself or toward your colleague as there is a chance of liquid being shot out of the test tube and harm you or your colleague.
7. Do not heat the standard glassware because its size will change.
8. Mix the chemicals inside the test tube well either by shaking or by a stirring rod.
9. Many chemicals are toxic, so wash your hands before leaving the laboratory.

REAGENTS AND HOW TO USE THEM

Conducting laboratory experiments depends on the use of various reagents and chemicals, and there is always a laboratory in a specified amount of reagents, part of which is placed on the laboratory benches and store the other in the cabinets that are designed for this purpose.

It is important to know the compositions of the reagents and their main properties, flammability, toxicity and ability to form explosive mixtures with other reagents [2].

Solutions and reagents are kept in glass bottles sealed with glass, rubber or corkscrew stoppers. A piece of paper is affixed to each bottle containing the name and concentration of the reagent [2, 3].

The following primary rules are considered when reagents are used:

1. If there is no clear definition of the amount of reagents to be taken, take as little as possible of reagents and this will give better results, save time and reduce consumption and waste materials.
2. It is strictly forbidden to return the excess quantity of reagent that was taken to the bottle of the reagent in order to prevent contamination and the complete destruction of the reagent.
3. Used reagents must be returned to their places after usage.
4. Do not mix the stoppers of different reagent bottles as well as rubber pipetting bulbs between the pipettes used to take the reagents.
5. Special attention should be paid when dealing with toxic, harmful or flammable substances such as barium salts, mercury, arsenic, Carbon disulphide and others.

HOW TO WRITE LABORATORY REPORTS

Writing laboratory reports is an important part of the practical sessions and when writing reports, consider the following [1]:

1. The student must be equipped with a notebook to write her/his observations about the practical lesson first hand.
2. The student should submit a report of the experiment, explaining the following:
   a. Name of the experiment.
   b. The purpose of the experiment.
   c. The practical idea is based on experimentation with the explanation of the equations of interaction.
   d. Conclusion.
   e. Difficulties that occurred when conducting the experiment.
   f. Reasons for the failure of the experiment (if a failure occurs).

THE WORK TECHNIQUES IN THE LABORATORY

The work in the descriptive and quantitative analysis laboratories requires the identification of some general processes, the basis of this practice and its mastery before conducting the analytical reactions required by the nature of the work to be performed. The student is supposed to have practiced these operations in advance, in addition, the student should know the following [3]:

1. Ensure the validity of the equipment and apparatus before usage with washing and cleaning them.
2. Preparation of solutions with required concentrations and indicators is done in the required analyses.
3. Precipitation and filtration of all kinds of aggregation, separation of precipitate away from supernatant and collection and maintenance of the precipitate.
4. The use of simple tools in work, for example, wash bottle, dropper and others, the technique of working in the laboratories, includes:

a) Apparatus used in the Chemical Analysis

It is preferable to use small amounts of chemicals in the chemical analysis as this is economical in the consumption of these chemicals and the detection takes a shorter time. Table **1.1** lists common apparatus that are used in the chemical laboratory. Apparatus includes the following:

1. *Test Tubes and Centrifuge Tubes*: The use of circular-end pipes for the testing and shroud type of Pyrex (heat resistant) in the case of heating.
2. *Droppers*: used to transferring solutions and reagents from bottles; there are different types of droppers.
3. *Reagents Bottles*.
4. *Centrifuge*: To separate the precipitant from the solution. The solution could be placed either in a test tube or a centrifuge tube in one of the spaces in the rotor and in the opposite space, place a balancing tube, *i.e.*, a tube of the same size filled with an equal volume of water to that of your solution.

The Benefit of Using Centrifuge:
   a. Speed and accuracy.
   b. To concentrate on the precipitate to a small volume can be quantified.
   c. Quickly and efficiently wash the precipitate.

5. *Wash Bottles*, *i.e.*, filled with distilled water: it is used by pressing it with hands [4].

b) Heating Solutions

Do not heat the centrifuge tube on the fire directly so that it could affect the content of the tube and cause serious damage, especially in the case of acids and bases. It is advisable to heat it on a water bath. Do not heat the ordinary glass on a Bunsen burner or other heat source because it breaks when exposed to heat and uses glass made of Pyrex or porcelain.

c) Dissolving the Precipitate

The reagent is added to the precipitate and heated in a water bath until the precipitate dissolves completely or partially. In the latter case, the precipitate is separated from the solution by the insoluble force.

d) Cleaning Glassware

The laboratory glassware is cleaned by washing them firstly with detergent and tap water followed, then rinsed with distilled or deionized water, then dried with a hydrochloric acid solution (2%) or dichromate solution (10%) and 10 mL concentrated sulphuric acid. To remove hard precipitates from dirty glassware, the nitric acid solution can be used [2, 3].

Table 1.1. A list of commonly used apparatus in the chemical laboratory.

Apparatus	
Spatula	Pipettes
Wash bottle	Test tube
Beakers	Centrifuge test tube
Watch glass	Filter paper
Funnel	Litmus paper
Stirring road	Balance
Tong	Reagent bottles
Conical flask	Droppers
Graduated cylinder	Clamp
Volumetric flasks	Ruck
Burette	Stand

Warning (1): When the acid is diluted with water, the acid should always be added to water dropwise and not the reverse with the continuous stirring of the mixture, especially in the case of dilatation of the sulfuric acid to avoid the acid volatilization.

Warning (2): It is preferable to use small cylinders to measure the amount of acid or concentrated bases. Use rubber pipetting bulbs with a pipette. Do not ever use your mouth to pull the liquid into a pipette.

Warning (3): Mixing chemicals well inside a test tube is done either by shaking the tube or using a glass rod.

CHAPTER 2

Inorganic Qualitative Analysis

Abstract: In this chapter, the different types of analysis are explained. In addition, the importance of the rule of solubility of compounds in water in separation and identification of the cations is emphasised.

Keywords: Qualitative analysis, Quantitative analysis, Solubility rules, Soluble compounds.

INORGANIC QUALITATIVE ANALYSIS

A number of the chemical analysis are used to assign the components of any unknown substance. It could be determining the components and quantitative ratios of those unknown substances. This chemical analysis will be briefly introduced below [4]:

Firstly: Modern Analysis Methods

Modern analysis methods that are based on the progress and development of science and technology have been achieved in order to design and construct high-sensitivity devices. The working principle of that device is to measure one and more of the physical properties of substances underinvesting that link to components and quantitative ratios of any unknown substance. Photoanalysis, electrochemical analysis or voltage analysis have been used so far.

Secondly: Traditional Analysis Methods

1. *Qualitative Analysis*

It is defined as a set of tests or chemical processes aimed to identify the material whether unknown material is a pure substance or mixture in a solid state or in a solution without paying attention to the weights of these components or the elements involved in the structure or the proportions of their existence. Therefore, the descriptive analysis is never affected by the loss of the sample for a fraction of its quantity as long as the purpose of this analysis is to determine the components of the relative presence or weights in the sample [2, 4].

2. Quantitative Analysis

It is used to determine the weights of the components or elements involved in the composition of the mixture or the chemical compound and its proportions. Therefore, the presence of impurities in the material affects the determination of these weights and the choice of analysis method that requires knowledge of the elements that made up the substance under investigation.

A number of producers are required precede to the quantitative analysis:

a) Conduct qualitative analysis to identify the elements and components involved in the composition of the substance and the impurities that contain.

b) Purify the material to be analyzed from the impurities before assigning the proportions of its components.

Scales Using in Qualitative Analysis

The qualitative analysis can be carried out in different types of scales, as summarized in Table **2.1** [4].

Table 2.1. The types of measuring scale used in the qualitative analysis.

Scale Qualitative Analysis	Macro	Micro	Semi-micro
Sample size	5-100 mL 0.5 – 1 g	≥ 1 mL 0.005 – 0.01 g	One drop (no more than 1 mL) 0.05 – 0.01 g

There is no real difference between micro qualitative analysis and semi-micro qualitative analysis, which is the result of the difference in the quantities used in the two cases. The difference is confined to the quantities and small volumes of the sediments and solutions that are used, as well as the use of the type of special tools for each analysis.

There is no advantage in the sensitivity of the semi-micro qualitative analysis. However, there are a number of advantages in another aspect, for example, saving time and chemicals. Filtration, sediment washing, and vaporization processes can take about an hour. These processes can be performed in a few minutes if the semi-micro-method is followed, thus allowing the student to analyse more samples on time and gain broader training in analytical methods [2, 4].

GLOSSARY IN QUALITATIVE ANALYSIS

Centrifuge: It is a device used to accelerate the stability of the sediment, making use of centrifugal force.

Suspension: It is the separation of liquid from stable sediment.

Filtration: It is the separation of a precipitate from a fluid through a medium. A liquid that passes through the filter called filtrate or filtered liquid, while the residue left on the filter paper is called filtered solid (separation processes are often done in semi-micro-analysis by centrifuge and circulation).

Sedimentation: It is the formation of an insoluble solid that forms when reagent reacts with a solution. The insoluble solid is called a precipitate.

Reagent: It a substance used to make a chemical change in the material under test, for example, hydrochloric acid is a reagent added to the solution of silver nitrate because it is a precipitate of silver chloride.

Turbidity: It is the occurrence of something like clouds as a result of small particles stuck in a solvent.

Washing: It is the cleaning of precipitate (usually with water) to remove the suspension of the solution from which it precipitates.

General Rules in the Solubility of Inorganic Compounds in water Inorganic compounds are classified based on their solubility in water into three groups [6, 7]:

i.   Soluble compounds: Soluble in water more than 10 g/L.
ii. Low solubility compounds: The water solubility of less than 1 g/L.
iii. Moderately soluble compounds: Compounds that are soluble in water more than 1 g/L and less than 10 g/L.

RULES OF SOLUBILITY OF INORGANIC COMPOUNDS

As the solubility of matter in water is important to be taken into consideration in the analysis, the student must be familiar with the following rules of solubility [5, 6]:

1. All salts of hydrochloric acid (chlorides) and hydro-bromic (bromides) are soluble in water as well as hydro-iodic acid (iodides) other than salts associated with Ag^+, Hg_2^{2+}, Pb^{2+} [lead (II) chloride and lead (II) bromide are mild soluble in cold water and dissolved in hot water].
2. All nitrate NO_3^-, nitrite NO_2^- and acetic salts are soluble in water.
3. All sulphate salts are dissolved in water except barium sulphate, strontium sulphate and lead (II) sulphate ($BaSO_4$, $SrSO_4$ and $PbSO_4$), which are mild soluble in water.
4. All sodium, potassium and ammonium salts are soluble in water except for some complex salts.
5. All oxides and hydroxides are poorly soluble in water except oxides and hydroxides of alkali and alkaline earth metals, ammonium hydroxide and barium hydroxide, strontium hydroxide and calcium hydroxide, which are mild soluble in water.
6. All silicates are insoluble in water except alkaline metal silicate.
7. All cyanides are insoluble in water except alkaline cyanides and mercury (II) cyanide.
8. All sulphide salts (salts derived from H_2S) are poorly soluble in water except alkali metal sulphide, ammonium sulphide and alkaline earth metal sulphide.
9. All carbonates, phosphates, urea, oxalates, chromates, resins, iron, and cyanides are either insoluble or low-soluble except for the alkali and ammonium metals, but they readily dissolve in the achieved acids. The carbonates and acid phosphates such as $Ca(HCO_3)_2$, $Ca(H_2PO_4)_2$ and others are dissolved in water.

GROUP SEPARATION AND IDENTIFICATION OF THE CATIONS

A small number of carbonates can be detected in the presence of a few other cations. For example, adding hydrochloric acid to a solution containing all cations, only the lead, silver and mercury are precipitated, because all other chlorides are soluble (see solubility rules). Then lead, silver and mercurous (Mercury I) are a group of ions that can be precipitated in the solution by adding the group reagent (HCl).

After separating this group of ions, another group can be precipitated by adding the reagent of this group.

There are 28 radical cations that will be separated during the detection and investigation of the identity of the unknown cation and are divided into five major groups depending on the variation in some of their chemical properties. There are four groups by special regrant in certain circumstances, while the fifth group is in the form of a solution, which led to the name of the soluble group.

There are principles used to explain the sequential steps and the conditions that enhance the group precipitate are shown in Flow chart **2.1** since all cations are present in the solution together as a mixture.

a. All cations in solution are transformed by reacting with hydrochloric acid into chlorides in an acidic medium. Cations that do not dissolve their chlorides in water are the first group chlorides consisting of silver Ag^+, lead Pb^{+2} and mercurous Hg_2^{2+}, unlike chlorides of other dissolved groups in water.

b. The cations of the second group containing (Bi^{3+}, Cd^{2+}, Cu^{2+}, Hg^{2+}, pb^{2+}, Sn^{4+}, Sn^{2+}, Sb^{3+}, As^{3+}) are precipitated from their dissolved chloride solution as sulphides using H_2S directly or indirectly (as a thioacetamide analysis that gives H_2S). Hydrochloric acid is used to control the concentration of H_3O^+ (0.2-0.3 M). It is important to know the concentration of sulphide ion that comes from H_2S which is enough to precipitate the second group cations leaving other cations in another group (3-5) that are soluble in solution.

c. The third group contains (Co^{2+}, Cr^{3+}, Fe^{2+}, Fe^{+3}, Al^{3+}, Zn^{2+}, Mn^{2+}, Ni^{2+}) which are separated after precipitation as a mixture of insoluble hydroxides and sulphides using two precipitators: ammonia, H_2S or thioacetamide in a moderate basic solution consisting of ammonium hydroxide and ammonium chloride, where ions (Zn^{2+}, Co^{2+}, Fe^{2+}, Ni^{2+}, Mn^{2+}) are precipitated as sulphides while ions (Al^{3+}, Fe^{3+}, Cr^{3+}) are precipitated in the form of hydroxides.

d. The fourth group consists of ions (Calcium Ca^{2+}, Strontium Sr^{2+}, and Barium Ba^{2+}). These ions are within a single group in the periodic table. Therefore, they are similar in their chemical reactions and their valence electrons. Their components are reductive agents, their ions appear only slightly inclined to form complexities and they do not degrade in water. Among the ions (Ca^{2+}, Sr^{2+}, Ba^{2+}). It is difficult to make a good separation between them. The ions of this group are separated from the fifth group in one of the two ways:

i.   The phosphate ion PO_4^{3-} is added as a precipitating agent in a high concentration of ammonia solution as a consequence of that (Ca^{2+}, Sr^{2+}, Ba^{2+}), ions are precipitated in a general phosphate form ($M_3(PO_4)_2$) while magnesium is precipitated as Ammonium magnesium phosphate $NH_4MgPO_4.6H_2O$.

ii. The carbonate ion CO_3^{2-} is added as a precipitating agent to the group ions and then the magnesium ion will be the fifth group (v) ammonium NH_4^+, sodium Na^+, potassium K^+, magnesium Mg^{2+}. The cations of this group are not precipitated in any reagent, but they can be precipitated by some specific reagent.

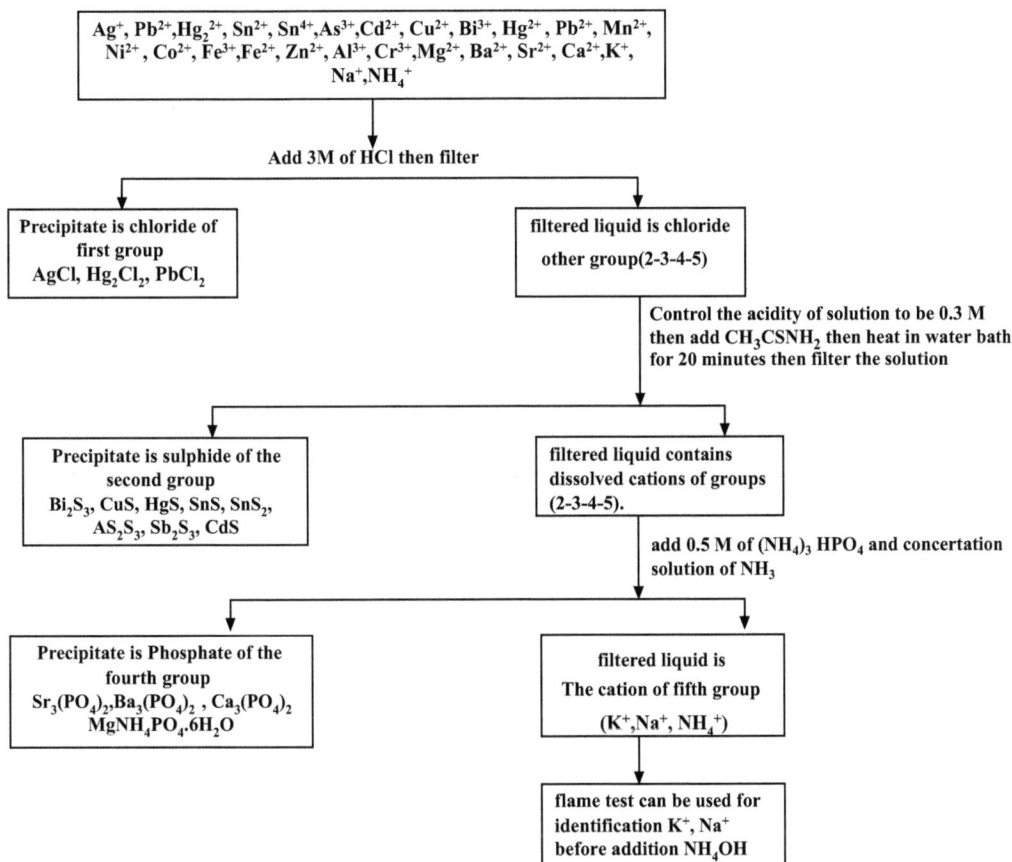

```
┌─────────────────────────────────────────────────────────────────┐
│ Ag⁺, Pb²⁺,Hg₂²⁺, Sn²⁺, Sn⁴⁺,As³⁺,Cd²⁺, Cu²⁺, Bi³⁺, Hg²⁺, Pb²⁺, Mn²⁺, │
│ Ni²⁺, Co²⁺, Fe³⁺,Fe²⁺, Zn²⁺, Al³⁺, Cr³⁺,Mg²⁺, Ba²⁺, Sr²⁺, Ca²⁺,K⁺,    │
│                          Na⁺,NH₄⁺                                 │
└─────────────────────────────────────────────────────────────────┘
```

Add 3M of HCl then filter

| Precipitate is chloride of first group $AgCl$, Hg_2Cl_2, $PbCl_2$ | filtered liquid is chloride other group(2-3-4-5) |

Control the acidity of solution to be 0.3 M then add CH_3CSNH_2 then heat in water bath for 20 minutes then filter the solution

| Precipitate is sulphide of the second group Bi_2S_3, CuS, HgS, SnS, SnS_2, AS_2S_3, Sb_2S_3, CdS | filtered liquid contains dissolved cations of groups (2-3-4-5). |

add 0.5 M of $(NH_4)_3 HPO_4$ and concertation solution of NH_3

| Precipitate is Phosphate of the fourth group $Sr_3(PO_4)_2$,$Ba_3(PO_4)_2$, $Ca_3(PO_4)_2$ $MgNH_4PO_4.6H_2O$ | filtered liquid is The cation of fifth group (K^+,Na^+, NH_4^+) |

flame test can be used for identification K^+, Na^+ before addition NH_4OH

Flow chart (2.1). The analysis for cations separation.

The following Tables (**2.2-2.7**) shows the quantities, concentrations, and types of materials used in each group from I to VI:

Table 2.2. Cations group I: Hg_2^{2+} mercury(I) ion, Pb^{2+} lead ion, Ag^+ silver ion.

Reagent	Concentration	Amount
$AgNO_3$	0.1N	250 mL
$Hg_2(NO_3)_2$	0.1N	250 mL
$Pb(NO_3)_2$	0.1N	500 mL
HCl	2 N	500 mL
K_2CrO_4	0.5N	250 mL
NH_3	30%	-

Table 2.3. Cations group II: Cu^{2+} copper ion, Cd^{2+} Cadmium ion, Hg^{2+} mercury(II) ion, Pb^{2+} lead ion.

Reagent	Concentration	Amount
$Cu(NO_3)_2.3H_2O$	0.1M	250 mL
$Cd(NO_3)_2$	0.1M	250 mL
$Hg(NO_3)_2$	0.1M	250 mL
C_2H_5NS	0.1 M	500 mL
HCl	0.3 M	250 mL
$K_4Fe(CN)_6$	0.3 M	250 mL
KCN	0.3 M	250 mL
$SnCl_2$	0.1 M	100 mL
HNO_3	3 M	250 mL
CH_3COOH	0.1 M	250 mL
NH_3	30%	-
0.1 mg $C_{24}H_{28}N_3Cl$ in 100 mL in DW		

Table 2.4. Cations group III: Fe^{3+} iron(III) ion, Al^{3+} aluminum ion, Cr^{3+} chromium(III) ion.

Reagent	Concentration	Amount
$FeCl_3$	0.1N	250 mL
$AlCl_3$	0.1N	250 mL
$Cr(NO_3)_3$	0.1M	250 mL
C_2H_5NS	0.1 M	500 mL
HCl	2 N	500 mL
NaOH	4 N	250mL
KCN	0.1 M	250 mL
$K_4[Fe(CN)_6]$	0.1 M	250 mL
$Pb(CH_3COO)_2$	0.1N	250 mL

(Table 2.4) cont.....

HNO$_3$	3 M	250 mL
CH$_3$COOH	0.1 N	250 mL
NH$_3$	30%	-
HCl	30%	HCl
H$_2$O$_2$	3%	H$_2$O$_2$
NH$_4$Cl	-	-
NH$_4$NO$_3$	-	-
C$_5$H$_{12}$O	-	-

Table 2.5. Cations group IV: Ni^{2+} nickel ion, Co^{2+} cobalt(II) ion, Mn^{2+} manganese(II) ion, Zn^{2+} zinc ion.

Reagent	Concentration	Amount
Ni(NO$_3$)$_2$.6 H$_2$O	0.1N	250 mL
Co(NO$_3$)$_2$.6 H$_2$O	0.1N	250 mL
MnCl$_2$	0.1N	250 mL
ZnCl$_2$	0.1 N	250 mL
HCl	2 N	500 mL
NaOH	3 M	250 mL
C$_2$H$_5$NS	0.1 M	500 mL
NH$_3$	30%	-
H$_2$O$_2$	3%	-
1mg C$_4$H$_8$N$_2$O$_2$ in 100 mL ethanol		
NaF	-	-
Pb$_3$O$_4$	-	-
NH$_4$Cl	-	-
NH$_4$SCN	-	-
C$_5$H$_{12}$O	-	-

Table 2.6. Cations group V: Ba^{2+} barium ion, Sr^{2+} strontium ion, Ca^{2+} calcium ion.

Reagent	Concentration	Amount
Ca(NO$_3$)$_2$	0.1N	250 mL
Sr(NO$_3$)$_2$	0.1N	250 mL
Ba(NO$_3$)$_2$	0.1N	250 mL
HCl	12 M	-
NH$_4$OH	6 M	-
(NH$_4$)$_2$CO$_3$	0.1 M	-
CH$_3$COONH$_4$	0.1 M	-
K$_2$CrO$_4$	0.1 M	-
CH$_3$COOH	6M	-

Table 2.7. Cations group VI: NH_4^+ ammonium ion, Mg^{2+} magnesium ion, K^+ potassium ion, Na^+ sodium ion.

Reagent	Concentration	Amount
$Mg(NO_3)_2$	0.1N	250 mL
NH_4NO_3	0.1N	250 mL
KNO_3	0.1N	250 mL
HNO_3	0.1 N	250 mL
NaOH	6 M	-
NH_4OH	6 M	-
Na_2HPO_4	0.1	-
CH_3COOH	12M	-
$Na(C_6H_5)_4B$	-	-

CHAPTER 3

The First Group

Abstract: Chapter three provides the procedure for the detection and separation of the first group cations, highlighting the importance of the solubility constant and the effect of pH on the detection and separation of the cations.

Keywords: Chlorides, Detection, Separation, Silver group, The solubility constant ksp.

DETECTION AND SEPARATION OF THE FIRST GROUP OF CATIONS

A number of the chemical analysis are used to assign the components of any unknown substance. It could be determining the components and quantitative ratios of those unknown substances. This chemical analysis will be briefly introduced below [4].

Silver Group

The first group of cations (silver group) consists of three cations, Ag^+ silver ion, Pb^{2+} lead ion, Hg_2^{2+} mercurous ion, and their reagents are diluted in hydrochloric acid because the chlorides of previous cations are insoluble in water ($AgCl$, Hg_2Cl_2) or mildly soluble in water ($PbCl_2$) [5].

While the cations of chlorides of the other five groups are dissolved in water, the table below shows that the solubility of the mercurous chloride exhibits the lowest value.

Lead chloride has a significant solubility in water, therefore, it is not completely precipitated and separated in the first group at a room temperature of 25°C. Thus, lead is present in the second group and is precipitated as lead sulphide (Table **3.1**), a list of Solubility constant ksp at 25°C various ionic compounds in water at 25°C is given in Appendices [6, 7].

Table 3.1. The solubility constant ksp for chloride; the first group cation at 25°C.

Compounds	Solubility Constant k_{sp} at 25°C	Dissolved Amount mg/L at 25°C
AgCl	1.8×10^{-10}	1.7×10^{-3}
PbCl$_2$	1.7×10^{-5}	1.6×10^{-2}
Hg$_2$Cl$_2$	1.3×10^{-18}	3.8×10^{-4}

Chloride ion obtained from the dissociation of hydrochloric acid in the solution is added to cations of the first group, leading to complete precipitation of Ag^+, Pb^{2+} as a result of common ions. However, adding chloride abundantly would lead to a good amount of HCl for dissolving some of the Ag^+ and Pb^{2+} and as a result of which the complex ion such as $[AgCl_2]^-$, $[PbCl_3]^-$ that are soluble are formed. For this reason, the reagent of this group is hydrochloric acid achieved according to the following reactions:

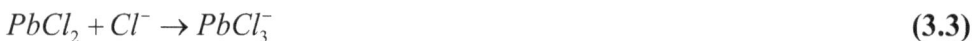

$$AgCl + Cl^- \rightarrow AgCl_2^- \tag{3.1}$$

$$AgCl + 2Cl^- \rightarrow AgCl_3^{2-} \tag{3.2}$$

$$PbCl_2 + Cl^- \rightarrow PbCl_3^- \tag{3.3}$$

Separation of the Cations of Group I

The first group contains the following cations: Hg_2^{2+}, Pb^{2+}, Ag^+

1. Take 1 mL of a solution containing the first group cations and place it in a 5mL conical tube.
2. Add about 15 drops of diluted HCl, shake the solution well, the cations of the first group will be precipitated because their chlorides are insoluble as in the following equations [3, 5]:

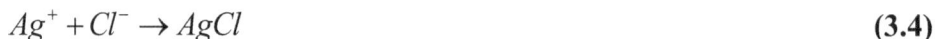

$$Ag^+ + Cl^- \rightarrow AgCl \tag{3.4}$$

White precipitate

$$Hg_2^{2+} + 2Cl^- \rightarrow Hg_2Cl_2 \qquad\qquad (3.5)$$

White precipitate

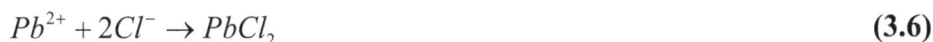

$$Pb^{2+} + 2Cl^- \rightarrow PbCl_2 \qquad\qquad (3.6)$$

White precipitate

3. Place the tube in a centrifuge for 5 minutes, make sure that full precipitation is obtained for the cations by adding a drop of HCl.
4. Filter the solution, the filtered liquid contains the cations from other groups (2, 3, 4, 5 and 6) that is neglected and the precipitate is a mixture of AgCl, Hg_2Cl_2 $PbCl_2$.
5. Add 15 drops of hot water to the precipitate (the precipitate of lead chloride dissolve in hot water with stirring). Use the centrifuge to separate the mixture for 5 minutes. The precipitate contains AgCl, Hg_2Cl_2 and the filtered liquid contains Pb^{2+}. Lead is identified by depositing it in the form of yellow lead chromate.

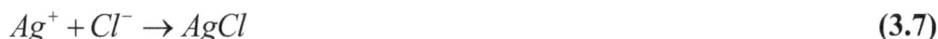

$$Ag^+ + Cl^- \rightarrow AgCl \qquad\qquad (3.7)$$

White precipitate

6. The precipitate contains AgCl. Hg_2Cl_2 is separated by ammonia solution and 15 drops of NH_4OH to the precipitate is added, as a consequence of that, the silver chloride is dissolved in the ammonia solution by formation of the complex $[Ag(NH_3)_2]Cl$ silver diamine chloride.

$$AgCl_{(s)} + 2NH_4OH \rightarrow Ag(NH_3)_2^+ + Cl^- + H_2O \qquad\qquad (3.8)$$

Soluble complex

The filtered liquid contains Ag^+, which is detected by the addition of 5 drops of diluted HCl. A white solid is formed to indicate the presence of silver.

$$Ag(NH_3)_2^+ + 2HCl \rightarrow AgCl_{(s)} + 2NH_4^+ \qquad\qquad (3.9)$$

As for the separated precipitate, it is caused by the oxidation of mercurous ion to mercuric ion in the presence of ammonia, where part of the mercuric ion is reduced to mercury (black precipitate) and the other part is a white precipitate that forms mercury amide chloride [1].

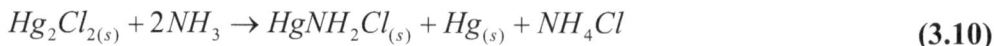

$$Hg_2Cl_{2(s)} + 2NH_3 \rightarrow HgNH_2Cl_{(s)} + Hg_{(s)} + NH_4Cl \qquad\qquad \textbf{(3.10)}$$

Mercuric amidochloride   Black metallic

Detection and separation of the first group of cations is summarised in Flow chart **3.1**.

ADDITIONAL INFORMATION:

1. A slight excess of HCl helps to accomplish precipitation, as it prevents precipitation of Bismuth oxychloride and Antimoine oxychloride, the large excess of HCl will lead to dissolve a part of the precipitate of silver chloride and lead (II) chloride [3].
2. Lead (II) chloride may dissolve in the washing process and, therefore, the water must contain HCl because the solubility of Lead (II) chloride is reduced due to the impact of a common ion.
3. Lead (II) chloride dissolves in hot water and precipitates with cold water.
4. In the case of a presence of lead cation in enough amount, Lead (II) may be precipitated with a precipitate of silver chloride, so any traces of Lead (II) chloride should be removed by washing with distilled water and then swashing mixture of distilled water with potassium chromate in order to detect, the presence of lead cation.

    The presence of mercury cation has an impact on dissolving the silver chloride in ammonium solution, mercury cation reduces silver cation to silver and the last one does not dissolve in the ammonium hydroxide. Therefore, the displacement reaction will occur and the silver cation could not detect; hence the cations should be separated quickly.

$$2Hg + 2Ag(NH_3)_2^+ \rightarrow Hg_2^{2+} + 2Ag_{(s)} + 4NH_3 \qquad\qquad \textbf{(3.11)}$$

5. Lead (II) can be precipitated by sulphuric acid in the form of sulphate, but it is better to precipitate in the form of chromate because Lead(II) chromate is precipitated first because it is less soluble than lead sulphate

$$Pb^{2+} + CrO_4^{2-} \rightarrow PbCrO_4 \qquad \mathbf{K_{Sp} = 2 \times 10^{-16}} \qquad (3.12)$$

$$Pb^{2+} + SO_4^{2-} \rightarrow PbSO_4 \qquad \mathbf{K_{Sp} = 1.3 \times 10^{-8}} \qquad (3.13)$$

6. Sodium chloride can be used as a precipitating agent rather than hydrochloric acid because it provides chloride anions, which works to precipitate the cations of this group in the form of chlorides [1].

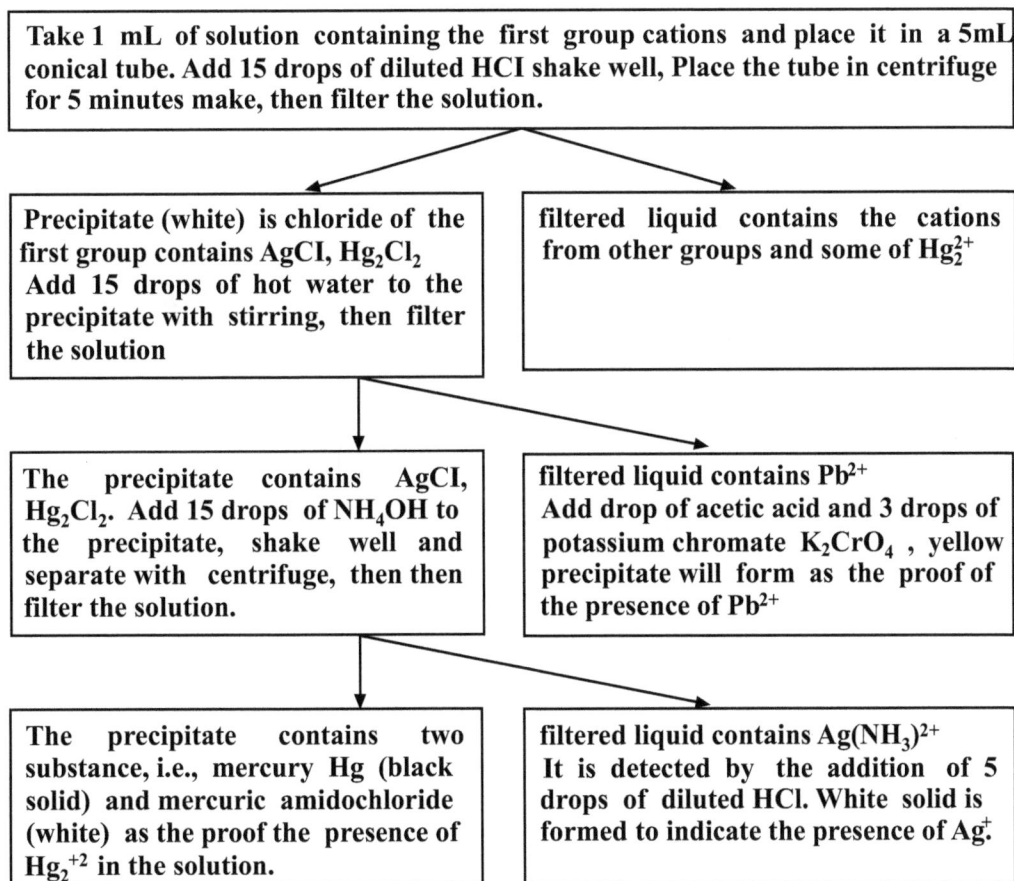

Take 1 mL of solution containing the first group cations and place it in a 5mL conical tube. Add 15 drops of diluted HCl shake well, Place the tube in centrifuge for 5 minutes make, then filter the solution.

Precipitate (white) is chloride of the first group contains AgCl, Hg_2Cl_2
Add 15 drops of hot water to the precipitate with stirring, then filter the solution

---

filtered liquid contains the cations from other groups and some of Hg_2^{2+}

The precipitate contains AgCl, Hg_2Cl_2. Add 15 drops of NH_4OH to the precipitate, shake well and separate with centrifuge, then then filter the solution.

---

filtered liquid contains Pb^{2+}
Add drop of acetic acid and 3 drops of potassium chromate K_2CrO_4 , yellow precipitate will form as the proof of the presence of Pb^{2+}

The precipitate contains two substance, i.e., mercury Hg (black solid) and mercuric amidochloride (white) as the proof the presence of Hg_2^{+2} in the solution.

---

filtered liquid contains $Ag(NH_3)^{2+}$
It is detected by the addition of 5 drops of diluted HCl. White solid is formed to indicate the presence of Ag^+.

Flow chart (3.1). The analysis of cations group I.

CHAPTER 4

The Second Group

Abstract: This chapter discusses the detection and separation of the second group of cations. It gives step by step practical directions and a set of rules, which must be followed for successful detection and separation of the cations.

Keywords: Acidic medium, Common Ion, Sulphides, Thioacetamide.

DETECTION AND SEPARATION OF THE SECOND GROUP CATIONS

Copper Group

The group includes Mercury (II) Hg^{2+}, Copper Cu^{2+}, Lead Pb^{2+}, Bismuth Bi^{3+}, Cadmium Cd^{2+}, Tin (II) Sn^{2+}, Tin (IV) Sn^{4+}, Antimony (III) Sb^{3+}, Antimony (V) Sb^{5+} and Arsenic (III) As^{3+}, Arsenic (V) As^{5+}. This group is divided into two subgroups: the first group is called copper group, which contains Mercury Hg^{2+}, Copper Cu^{2+}, Lead Pb^{2+}, Bismuth Bi^{3+} and Cadmium Cd^{2+}, whereas the second family is called the arsenic group, which contains: Sn^{2+}, Sn^{4+}, Sb^{3+}, Sb^{5+}, As^{3+} and As^{5+}. The reason for this division is based on their solubility's of some its sulphides in ammonium polysulphide $(NH_4)_2S_x$; x>2 [5].

It is found that the copper family (CuS, PbS, CdS, Bi_2S_3 and HgS) does not dissolve in $(NH_4)_2S$, while, the arsenic family (SnS_2, SnS, Sb_2S_5, Sb_2S_3 and AsS_5, As_2S_3) is soluble in ammonium polysulphide.

The reagent of the second group is hydrosulfuric acid (H_2S) in the acidic medium (0.3 N of HCl) because the sulphide of the former elements does not dissolve in this concentration of acid.

It is important to understand the function of hydrosulfuric acid in descriptive analysis and the conditions in which acid is used in the precipitation of the second and fourth groups. The analyst should be aware of the reactions taking place in the test tube and the concepts of equilibrium and solubility product constant.

Hydrosulfuric acid H_2S is a diprotic acid and a weak acid. Its two stages of ionization are:

$$H_2S \rightarrow HS^- + H^+ \qquad \mathbf{K_1 = 1.0 \times 10^{-8}} \qquad\qquad (4.1)$$

$$HS^- \rightarrow S^{2-} + H^+ \qquad \mathbf{K_2 = 1.3 \times 10^{-15}} \qquad\qquad (4.2)$$

By combining these balances, we get the following balance:

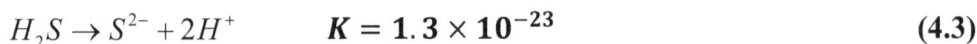

$$H_2S \rightarrow S^{2-} + 2H^+ \qquad \mathbf{K = 1.3 \times 10^{-23}} \qquad\qquad (4.3)$$

Which is equal to its second hitting the two sets k_2, k_1, *i.e.*, $(S^{2-})(H^+)^2 = (H_2S)$. When the solution is saturated, the concentration of H_2S is approximated (0.1) N, therefore, $(S^{2-})(H^+)^2 = 1.3 \times 10^{-23}$.

This relationship shows that the concentration of sulphide ions is inversely square of the concentration of hydrogen ion, which means that the sulphide (S^{2-}) concentration decreases when the acidity of the solution increases. The concentration of these ions can be controlled by adjusting the acidity of the solution. As the previous relationship shows, a specific acid concentration is used so that the second group is only precipitated. Therefore, the cations of the second group require a lower concentration of the S^{2-} ion to be precipitated in the comparison between the cation of the third and fourth group. In fact, the solubility of the second group sulphide, in general, was lower than the sulphides of the third and fourth groups. Table **4.1** shows the fixed solubility values of some of the first four groups of sulphides [6-8].

Table 4.1. Solubility constant values of some sulphides compounds.

Metal Sulphide	K_{sp}	Metal Sulphide	K_{sp}
CuS	6.3×10^{-36}	NiS	1.3×10^{-25}
CdS	8×10^{-27}	CoS	5×10^{-22}
PbS	8×10^{-28}	Sb_2S_3	$<10^{-30}$
HgS	4×10^{-53}	MnS	2.5×10^{-13}
As_2S_3	8×10^{-25}	FeS	6.3×10^{-18}
Bi_2S_3	1.6×10^{-72}	ZnS	1.6×10^{-24}
SnS	1×10^{-25}	Ag_2S	6.3×10^{-50}

The concentration of about 0.3 N hydrosulfuric acid was chosen, so the concentration of sulphide ion (S^{2-}) was sufficient for cadmium sulphide CdS to be precipitated, *i.e.*, cadmium sulphide is more soluble of the second group. At the same time, the concentration of sulphide ion (S^{2-}) should be less than required to precipitate zinc sulphide ZnS *i.e.*, zinc sulphide is the lowest solubility of the fourth group. Therefore, the acidity of the solution plays a significant factor in the precipitation of the second group and any error may lead to precipitate the group V [5, 9].

A Mercury (II), Copper (II), Cadmium (II) are located within the transitional elements in the periodic table; the other five elements are located in line between metal and nonmetal Tin (II) and IV, Antimony (III) and (V), Arsenic (III), Bismuth (III) and lead (II). As a result of that, these elements show amphoteric properties that differ in intensity between one element and another. The elements of this group are characterized by their strong tendency to bond to sulfur. This property appears in two forms as a precipitate or the form of metals in nature [1, 4].

The Method of Work

1. Take 1 mL of a solution containing the group ions, *i.e.*, a copper family that is insoluble in ammonium polysulphide (HgS, Bi_2S_3, PbS, CuS, CdS).
2. In a conical tube, then add a drop of the methyl violet indicator.
3. Add diluted hydrochloric acid HCl until the colour of the solution turns yellow-green, which means that the acid concentration is 0.3M.
4. Then, add hydrosulfuric acid H_2S 20 drop in the form of thioacetamide solution as seen in Equation (4.4), so the toxicity and disagreeable odor of hydrogen sulphide are eliminated [9].

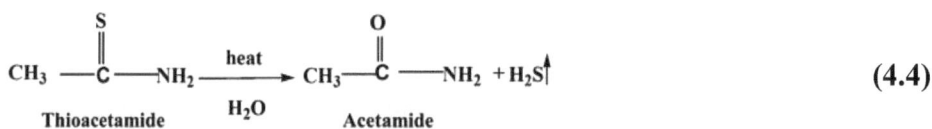

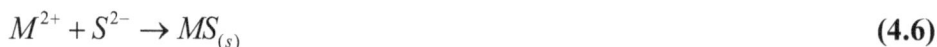

$$CH_3-\overset{\overset{S}{\|}}{C}-NH_2 \xrightarrow[\text{H}_2\text{O}]{\text{heat}} CH_3-\overset{\overset{O}{\|}}{C}-NH_2 + H_2S\uparrow \qquad (4.4)$$

Thioacetamide   Acetamide

$$Hg^{2+} + S^{2-} \rightarrow HgS_{(s)} \qquad (4.5)$$

$$M^{2+} + S^{2-} \rightarrow MS_{(s)} \qquad (4.6)$$

5. Heat the solution in a water bath for 20 minutes; copper family cations will be precipitated in the form of sulphide as Equation (4.6), then place a conical tube in a centrifuge for the separation for 5 minutes.

6. The filtrate contains other cations of groups (III, V, IV) while the precipitate contains HgS, Bi_2S_3, PbS, CdS and CuS. The ions are separated by adding 5 drops of diluted nitric acid followed by separation in a centrifuge for 5 minutes. All sulphides dissolved except mercury sulphide. Mercury sulphide is precipitated in the form of a black solid [5].

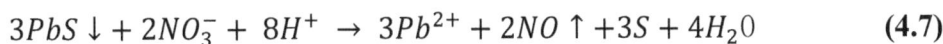

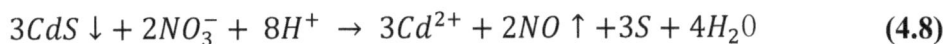

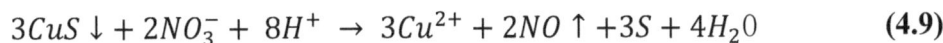

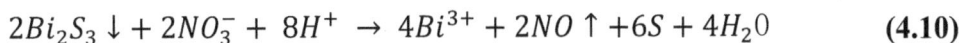

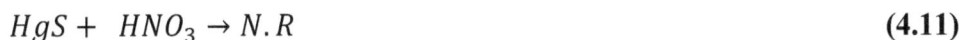

$$3PbS \downarrow + 2NO_3^- + 8H^+ \rightarrow 3Pb^{2+} + 2NO \uparrow +3S + 4H_2O \qquad (4.7)$$

$$3CdS \downarrow + 2NO_3^- + 8H^+ \rightarrow 3Cd^{2+} + 2NO \uparrow +3S + 4H_2O \qquad (4.8)$$

$$3CuS \downarrow + 2NO_3^- + 8H^+ \rightarrow 3Cu^{2+} + 2NO \uparrow +3S + 4H_2O \qquad (4.9)$$

$$2Bi_2S_3 \downarrow + 2NO_3^- + 8H^+ \rightarrow 4Bi^{3+} + 2NO \uparrow +6S + 4H_2O \qquad (4.10)$$

$$HgS + HNO_3 \rightarrow N.R \qquad (4.11)$$

7. The precipitate contains black HgS that dissolved in the Aqua Regia. It consists of a mixture of concentrated nitric acid and hydrochloric acid (HNO_3: HCl) (1:3). Aqua Regia takes its name as it could dissolve gold and other so-called noble metals.

8. To dissolve HgS: Add 6 drops of concentrated hydrochloric acid in a test tube, then add 2 drops of concentrated nitric acid and the mixture to HgS followed by 3 drops of Tin(II) chloride $SnCl_2$. The mercury ion will be detected through the reduction process of the mercury (II) $HgCl_2$ to the mercury Hg_2Cl_2 (I) and then to the mercury element (Hg) as shown in the following equations:

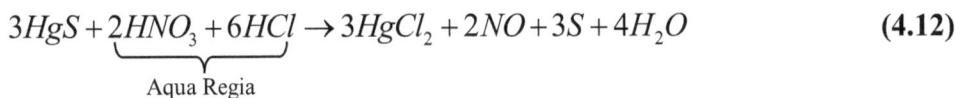

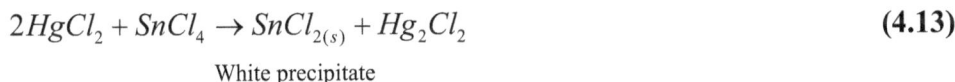

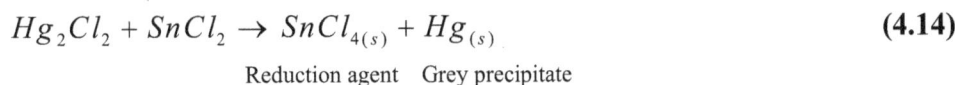

$$3HgS + \underbrace{2HNO_3 + 6HCl}_{\text{Aqua Regia}} \rightarrow 3HgCl_2 + 2NO + 3S + 4H_2O \qquad (4.12)$$

$$2HgCl_2 + SnCl_4 \rightarrow SnCl_{2(s)} + Hg_2Cl_2 \qquad (4.13)$$

White precipitate

$$Hg_2Cl_2 + SnCl_2 \rightarrow SnCl_{4(s)} + Hg_{(s)} \qquad (4.14)$$

Reduction agent   Grey precipitate

Other cations from group II (Cu, Pb, Bi, Cd) which are in the form of nitrate are separated by adding ammonia solution (NH_4OH) and 25 drops of ammonia. Pb^{2+} and Bi^{3+} are precipitated as the form of hydroxide, cadmium and copper melt and are converted into complex ions dissolved in the solution.

$$Pb(NO_3)_2 + 2NH_4OH \xrightarrow{\;2NH_3\;} Pb(OH)_{2(s)} \qquad\qquad (4.15)$$

White precipitate

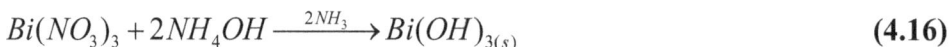

$$Bi(NO_3)_3 + 2NH_4OH \xrightarrow{\;2NH_3\;} Bi(OH)_{3(s)} \qquad\qquad (4.16)$$

$$Cu(NO_3)_2 + 2NH_4OH \xrightarrow{\;2NH_3\;} Cu(NH_3)_4(NO_3)_2 \qquad\qquad (4.17)$$

Dissolved complex

$$Cd(NO_3)_2 + 2NH_4OH \xrightarrow{\;2NH_3\;} Cd(NH_3)_4(NO_3)_2 \qquad\qquad (4.18)$$

Dissolved complex

9. Filtrate and precipitate are separated by a centrifuge for 5 minutes. The precipitate is dissolved in acetic acid CH_3COOH. First, the acidity of the solution is tested by a blue litmus paper, then 2 drops of potassium chromate K_2CrO_4 are added. A yellow solid is then precipitated as the evidence of the presence of Pd^{2+}.
10. The filtrate contains Cu^{2+}, Cd^{2+} and is divided into two parts. The first is to test the presence of copper by adding acetic acid until a blue litmus paper turns red, then 2-3 drops of Potassium ferrocyanide $K_4[Fe(CN)_6]$ are added. A reddish brown solid of Copper ferricyanide is formed, which indicates the presence of copper.

$$Cu^{2+} + [Fe(CN)_6]^{4-} \rightarrow Cu_2[Fe(CN)_6]_{(s)} \qquad\qquad (4.19)$$

The second part is used to identify Cd^{2+}. As Cu^{2+} is present in the filtrate, Potassium cyanide KCN has to be added, as it forms a colourless complex with copper tetracyanocuprate (I) $[Cu (CN)_4]^3$. The copper is univalent in the complex. The cyanogen partly reacts to form cyanate and cyanide ions. The copper complex is more stable and does not form a precipitate with H_2S. Tetracyanocadmiate (II) $[Cd(CN)_4]^{2-}$ is formed as a yellow precipitate with H_2S.

15 drops of thioacetamide are added and then heated for 15 minutes. The obtained product was yellow, which indicated the presence of Cd^{2+}.

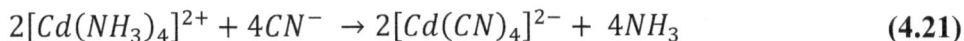

$$2[Cu(NH_3)_4]^{2+} + 10CN^- \rightarrow 2[Cu(CN)_4]^{3-} \downarrow +(CN)_2 + 8NH_3 \qquad (4.20)$$

$$2[Cd(NH_3)_4]^{2+} + 4CN^- \rightarrow 2[Cd(CN)_4]^{2-} + 4NH_3 \qquad (4.21)$$

$$(CN)_2 + 2NH_3 + H_2O \rightarrow CNO^- + CN^- + 2NH_4^+ \qquad (4.22)$$

$$2[Cd(CN)_4]^{2-} + H_2S + 2NH_3 \rightarrow CdS \downarrow +2NH_4^+ + 4CN^- \qquad (4.23)$$

Detection and separation of the second group cations are summarised in the Flow chart **4.1**.

ADDITIONAL INFORMATION: How to separate the cations of the second group [1].

1. The acidity of the solution should be 0.3 N, which is necessary to separate the cations of the second group. The increase in the acidity of the solution may block or inhibit the precipitate of some ions such as Sn^{4+}, Pb^{2+} and Cd^{3+}. When the acidity of the solution is less than 0.3 N, the cations of the third group will precipitate, for example, Zn^{2+}, Ni^{2+} and Co^{2+} in the form of sulphides. An indicator is used to control the acidity. For example, methyl violet indicated that the solution turns to yellow-green at the desired acidity.
2. It is necessary to get rid of the excess gas H$_2$S. It is oxidized with oxygen air converted into sulfate SO_4^{2-}, which precipitate some cations of group IV such Ba^{2+}.
3. Hydrochloric acid is used to control the concentration of S^{2-}. As it is observed previously, the concentration of S^{2-} depends on the concentration of H^+. The effect of the common ion is controlled by the addition of hydrochloric acid S^{2-} as follow

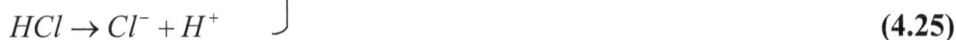

$$H_2S \rightleftarrows S^{2-} + 2H^+ \qquad (4.24)$$

$$\left.\begin{array}{l}\\ \\ \end{array}\right\} Common\ ions$$

$$HCl \rightarrow Cl^- + H^+ \qquad (4.25)$$

4. There are a number of acids that could not be used to adjust the pH of the solution, for example H$_2$SO$_4$, HNO$_3$ and CH$_3$COOH, because the use of H$_2$SO$_4$

leads to precipitate Pb^{2+} in the form of $PbSO_4$. As for using HNO_3, it dissolves some of the precipitates CdS, CuS, Bi_2S_3 and PbS as sulphides dissolve in nitric acid, *i.e.*, the oxidizing agent. As for CH_3COOH acid, it may lead to form a weakly ionized $Pb(CH_3COOH)_2$.

5. Nitric acid behaves as an oxidizing agent where S^{2-} is oxidized to sulfur and cation metal is composed expect mercury sulphide HgS, which is not affected by nitric acid. Nitric acid is used to dissolve other sulphides, for example, PbS, CuS, Bi_2S_3, CdS and is used to separate those sulphides from HgS.

6. Lead Hydroxide $Pb(OH)_2$ is amphoteric, which dissolves in acids and bases. This characteristic is used to separate $Pb(OH)_2$ and $Bi(OH)_3$ as $Bi(OH)_3$ does not dissolve in NaOH while $Pb(OH)_2$ dissolves in NaOH thereby, forming the dissolved complex ions PbO_2^{2-}.

The complex of $[Cu(CN)_4]^{3-}$ is more stable than the complex of $[Cd(CN)_4]^{2-}$, taking advantage of this characteristic, Cu^{2+} and Cd^{2+} are separated. Therefore, the concentration of Cu^{2+} is very low in the solution. The decomposition constant of the complex $[Cd(CN)_4]^{2-}$ $(5x10^{-28})$ is relatively high than the decomposition of the complex $[Cu(CN)_4]^{3-}$ $(1.4x10^{-17})$ as a result of that, the concentration of Cd^{2+} is sufficient to be precipitated when H_2S is introduced, although solubility product constants Ksp for CdS and CuS are $1x10^{-27}$ and $8x10^{-38}$ respectively.

7. In the case of the presence of Bi^{3+} and Pb^{2+} in the solution, it is important to adjust the acidity of the solution by acetic acid CH_3COOH before the addition of the CrO_4^{2-} into the solution when Pb^{2+} is detected as Bi^{3+} and may be precipitated. In addition, it is important to dilute a concentrated acid to avoid dissolving $PbCrO_4$ when there are high concentrations of H^+.

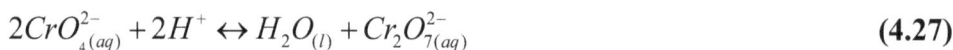

$$PbCrO_4 \leftrightarrow Pb^{2+} + CrO_4^{2-} \qquad\qquad (4.26)$$

$$2CrO_{4(aq)}^{2-} + 2H^+ \leftrightarrow H_2O_{(l)} + Cr_2O_{7(aq)}^{2-} \qquad\qquad (4.27)$$

Take 1 mL of solution containing (Cd^{2+}, Cu^{2+}, Bi^{3+} and Pb^{2+})and place it in a 5 mL conical tube. Add (1-2) drops of methyl violet indicator (MV), shake well, then add 20 drops from HCl. The solution turn green yellow. Place the tube in water bath for 20 minutes then place it in a centrifuge for 5 minutes then filter the solution.

Filtrate is negligent as it contains other ions of group cations

Precipitate contains (Bi_2S_3, HgS, PbS, CdS, CuS). Add 15 drops of diluted nitric acid, heat for 5 minutes followed by a separation in a centrifuge for 5 minutes then filter the solution.

Filtrate contains (Cu^{2+}, Pb^{2+}, Bi^{3+}, Cd^{2+}) adding 25 drops of ammonia. Then a separate in a centrifuge for 5 minutes

The precipitate contains black HgS that dissolved in the Agua Regia, add mixture of concentrated acids (2drops HNO_3 +6 drops HCl) then add 3 drops of $SnCl_2$. Black solid is formed as indications of mercury.

Filtrate contains cadmium and copper in form of complex ions dissolved $Cu(NH_3)_4(NO_3)$ and $Cd(NH_3)_4(NO_3)$ in the solution. The solution is blue as the presence of Cu^{2+} The solution is divided into two parts.

The precipitate contains Pb^{2+}, Bi^{3+} as form of hydroxide, Add 3 drops of CH_3COOH followed by addition of K_2CrO_4. A yellow solid is precipitated as the evidence of the presence of Pd^{2+}. separate in a centrifuge for 5 minutes then filter the solution. Precipitate contains Pd^{2+} and filtrate contains Bi^{3+}

Potassium cyanide KCN has to be added dropwise until the blue color is disappeared then 15 drops is of thioacetamide is added, and then heated for 15 minutes. A yellow formed as indication of Cd^{2+} presence.

3 drops of acetic acid is added then Potassium ferrocyanide $K_4[Fe(CN)_6]$. A reddish brown solid of Copper ferricyanide is formed which indicates the presence of copper.

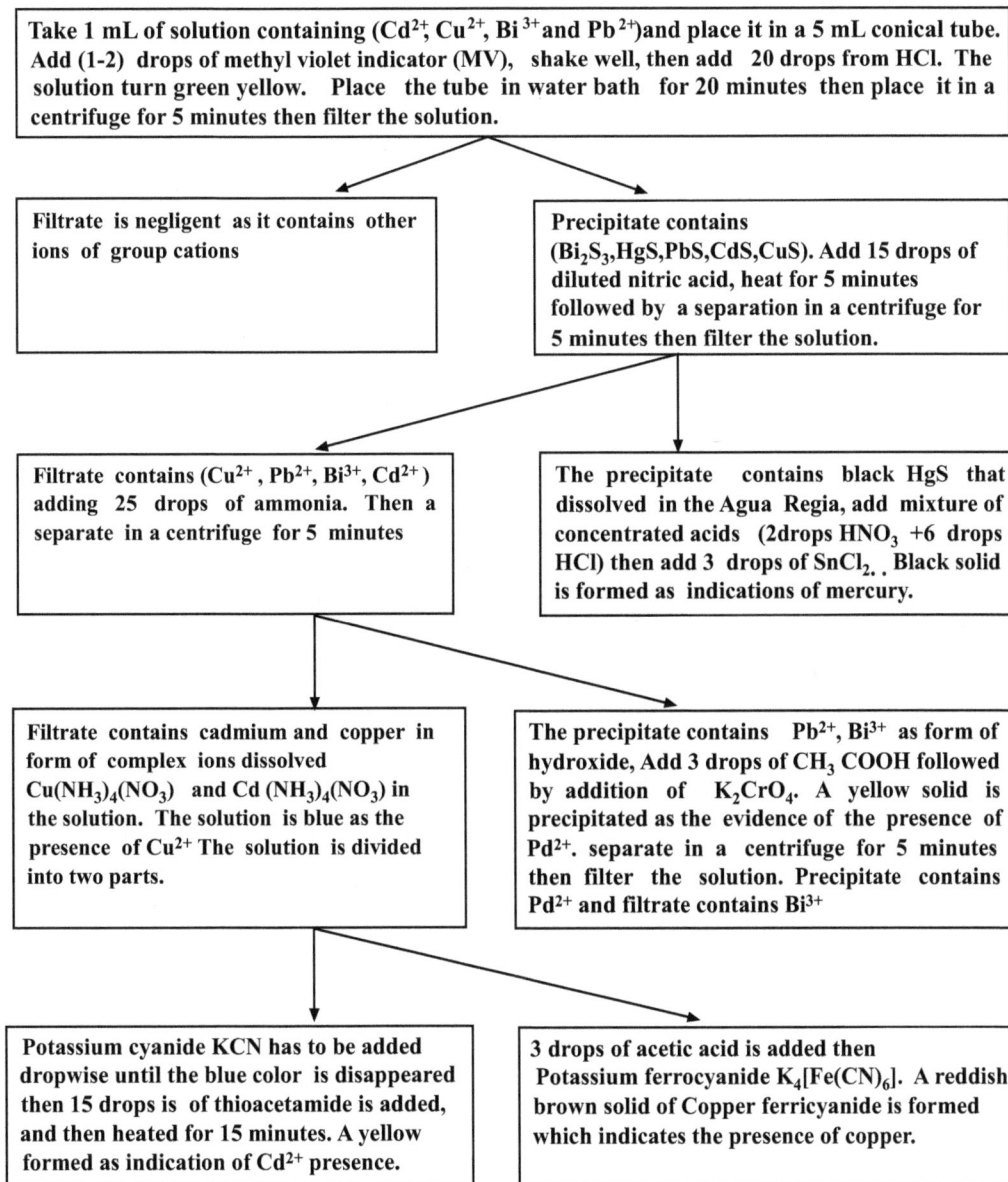

Flow chart (4.1). The procedure analysis for the copper group (II).

CHAPTER 5

The Third Group

Abstract: This chapter focuses on the third group cations (detection and separation) and describes the method of work, precipitate format, and the impact of pH on the procedure.

Keywords: Basic medium, Common ion, Hydroxides, Sulfides.

DETECTION AND SEPARATION OF THE THIRD GROUP CATIONS

Group III

This group includes the following cations: Iron (III and II) Fe^{3+} and Fe^{2+}, Aluminum Al^{3+}, Chromium Cr^{3+}, Manganese Mn^{2+}, Zinc Zn^{2+}, Nickel Ni^{2+}, Cobalt Co^{2+}. It can be divided into two subgroups depending on the type of precipitate as follows: (Fe^{3+}, Fe^{2+}, Al^{3+}, Cr^{3+}) are precipitated in the form of hydroxides. (Mn^{2+}, Zn^{2+}, Ni^{2+} and Co^{2+}) are precipitated as sulfides in a basic medium [5, 9].

This group (Fe^{3+}, Fe^{2+}, Al^{3+}, Cr^{3+}) contains Ammonium Hydroxide NH_4OH as the group reagent, the precipitates of this group are in the form of hydroxides because the hydroxides of these ions are insoluble in water, and the constants of solubility are low

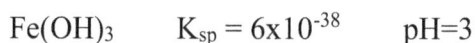

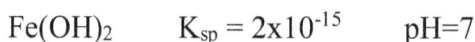

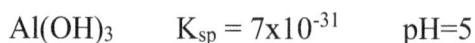

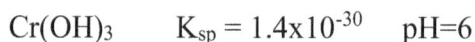

$Fe(OH)_3$	$K_{sp} = 6 \times 10^{-38}$	pH=3
$Fe(OH)_2$	$K_{sp} = 2 \times 10^{-15}$	pH=7
$Al(OH)_3$	$K_{sp} = 7 \times 10^{-31}$	pH=5
$Cr(OH)_3$	$K_{sp} = 1.4 \times 10^{-30}$	pH=6

The precipitation of metal hydroxide depends largely on the value of the hydroxide-soluble constant and the concentration of hydroxide ions in the solution or on the concentration of hydrogen ions (pH), where

Huda S. Alhasan & Nadiyah Alahmadi

$$pH = -\log\left[H^+\right] \qquad pOH = -\log\left[OH^-\right]$$

$$pK_a = -\log\left[K_a\right] \qquad pK_b = -\log\left[K_b\right]$$

$$pK_w = -\log\left[K_w\right] \qquad pK_w = pH + pOH = 14$$

Where K_w, K_a, and K_b are the autoionization water constant, the acid ionization constant, and base ionization constant, respectively. It could be controlled by the concentration of hydroxide by using a weak electrolyte, such as ammonium hydroxide (ammonia in water), and adding different concentrations of ammonium ions in the form of ammonium chloride salt. The added salt contains a common ion, *i.e.*, ammonium. The common ion is defined as the symmetric ion produced by two or more electrolyte ionization. If one electrolyte solution is added to another solution, one of the ions in the first solution leads to the concentration of sodium chloride to increase chloride ion concentration:

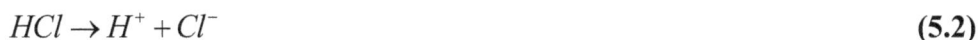

$$NaCl \rightarrow Na^+ + Cl^- \tag{5.1}$$

$$HCl \rightarrow H^+ + Cl^- \tag{5.2}$$

Adding ammonium chloride to the ammonia solution will increase the concentration of ammonium ion NH_4^+, but it will reduce the ionization of ammonia (the effect of the common ion and therefore the concentration of hydroxide ion)

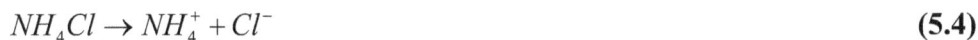

$$NH_3 + H_2O \leftrightarrow NH_4^+ + OH^- \tag{5.3}$$

$$NH_4Cl \rightarrow NH_4^+ + Cl^- \tag{5.4}$$

Controlling the concentration of hydroxide ions in the solution leads to precipitates of hydroxide and less soluble hydroxide.

Using a mixture of ammonium chloride and ammonia solution can precipitate selective ions. Some cations of the third group, for example, iron and chromium are located within the transition elements in the periodic table, therefore, these ions show the characteristics of the transition elements as follows:

1. Multi oxidations state as due to the proximity of the energy sublevels of 4p and 3d.
2. Coloured ions.
3. The tendency to form complex ions.

Aluminum is not from transition metal but it has similar properties to chromium and iron because these three have the same charge ions and the value of ionic radius is very close. The colors of the solution and precipitated ions are distinct, therefore the analyst should pay attention to these characteristics because they help a lot in predicting the identity of the solution, which leads to curtailing many of the necessary steps and processes.

Procedure

Group III is precipitated and separated from other groups and then their ions are detected in the mixture to be analyzed using the following steps [5, 9]:

1. Heat the filtered liquid that is produced by the separation of the second group to the boiling point until the evaporation of hydrogen sulfide gas is completely stopped.
2. Add 3-4 mL of concentrated nitric acid and continue boiling until all iron Fe^{2+}(II) is oxidized to iron Fe^{3+} (III) if present.
3. Add about 0.2 mg of solid ammonium chloride, then add the ammonia solution with stirring until the solution becomes basic and detected using the red litmus paper (until the smell of ammonia appears).
4. The previous solution is placed on a water bath for 3 minutes. After removing the solution from the water bath, the filtered liquid and solid precipitate are separated using the centrifuge device (C.F) for 5 minutes. This step is quick (step of separation), showing that the precipitate contains group III hydroxides. The color of the final formed precipitate was yellow-brown.

$$Al^{3+} + NH_4OH \xrightarrow{\ NH_4Cl\ } Al(OH)_{3(s)} \tag{5.5}$$

White precipitate

$$Fe^{3+} + NH_4OH \xrightarrow{\ NH_4Cl\ } Fe(OH)_{3(s)} \tag{5.6}$$

Brown precipitate

$$Cr^{3+} + NH_4OH \xrightarrow{\quad NH_4Cl \quad} Cr(OH)_{3(s)} \qquad\qquad (5.7)$$

Green precipitate

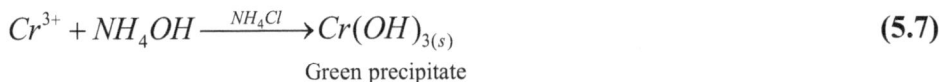

5. The solid residue that contains precipitate of this group is separated. The filtered liquid contains the ions of the subsequent groups. There is a possibility of the formation of a precipitate of the basic manganese oxide which belongs to group IV, especially if the filtration is not done immediately or if the ammonium chloride is not added in sufficient quantity.

6. 20 drops of NaOH solution and 10 drops of H_2O_2 (strong oxidizing agent) are used for precipitation then heated until boiling. Dissolved compounds of aluminum and chromium hydroxide will form, both are amphoteric (dissolved in acid and base). However, iron hydroxide is not dissolved. It forms a precipitate.

$$Fe(OH)_3 + Na\ OH \xrightarrow{\quad H_2O_2 \quad} Na(Fe(OH)_3)_{(s)} \qquad\qquad (5.8)$$

$$Al(OH)_3 + NaOH \xrightarrow{\quad H_2O_2 \quad} Na_2AlO_{2(aq)} + 2H_2O \qquad\qquad (5.9)$$

$$Cr(OH)_3 + NaOH \xrightarrow{\quad H_2O_2 \quad} Na_2CrO_{2(aq)} + 2H_2O \qquad\qquad (5.10)$$

7. The precipitate that contains iron is separated from the filtered liquid and divided into two parts, where the first part is dissolved in the diluted sulfuric acid and then in $K_4[Fe(CN_6)]$ Potassium Ferrocyanide. The formation of a blue precipitate indicates the presence of iron.

    The second part is dissolved in 5 drops of HNO_3 and then KSCN potassium thiocyanate solution is added. A red-brown precipitate of $[Fe\ (SCN)_6]^{3-}$ is formed.

    *The first part represents the reaction of Prussian blue, which is made up of ferro-ferric cyanide, which is a dark blue precipitate.

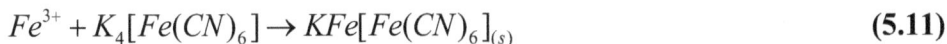

$$Fe^{3+} + K_4[Fe(CN)_6] \rightarrow KFe[Fe(CN)_6]_{(s)} \qquad\qquad (5.11)$$

Potassium Ferrocyanide　　　blue precipitate (Prussian blue)

The filtered liquid contains ions of Cr^{3+} and Al^{3+} and is divided into two parts:

The first part is used for the detection of Al^{3+} in several ways. The solution is acidified by HCl. Then, a mixture of NH_4Cl and NH_4OH is added, and $Al(OH)_3$ as a white precipitate is formed. After acidification of the solution, the solution is made basic by adding NaOH and then heated. A white precipitate of $Al(OH)_3$ is formed. The second part, Cr^{3+} ion, is detected in two ways:

1. If the filtered liquid is yellow, the solution is acidified by 6 M of acetic acid CH_3COOH. The acidity of the solution is tested by blue litmus paper. Then, Lead(II) acetate $Pb(CH_3COO)_2$ (0.1 M) is added to the solution. A yellow solid $PbCrO_4$ is precipitated indicating the presence of Cr^{3+}.
2. The filtered liquid is acidified by HNO_3 (strong oxidizing agent), then 8 drops of pentanol (Amyl alcohol) $C_5H_{11}OH$ is added. An organic layer will be formed, then 7 drops of H_2O_2 is added which initiates oxidation of Cr^{3+} to Cr^{6+}. A blue ring is formed in an organic layer of Chromic acid (H_2CrO_4), (see Flow chart **5.1**).

ADDITIONAL INFORMATION:

1. These ions (Cr^{3+}, Al^{3+}) have amphoteric properties, so their hydroxides dissolve in the concentrated NaOH, thus these ions (Cr^{3+}, Al^{3+}) are separated from Fe^{3+} as these are not dissolved in the basic medium.
2. When Fe^{2+} and Fe^{3+} are present in the solution, NH_4OH and NH_4Cl will precipitate only Fe^{3+} without Fe^{2+} because the concentration of OH^- is sufficient to precipitate Fe^{3+} because the concentration of OH^- is not sufficient to precipitate Fe^{2+}. See the pH value required for precipitate $Fe(OH)_2$.
   $Fe(OH)_2$ is precipitated in the absence of air. $Fe(OH)_2$ is quickly oxidized in case of exposure to the air and forms a red precipitate of $Fe(OH)_3$. It is also oxidized in the presence of H_2O_2 in the solution.

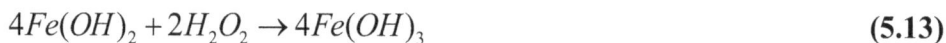

$$4Fe(OH)_2 + 2H_2O + O_2 \rightarrow 4Fe(OH)_3 \qquad\qquad (5.12)$$

$$4Fe(OH)_2 + 2H_2O_2 \rightarrow 4Fe(OH)_3 \qquad\qquad (5.13)$$

Cr^{3+} ion is precipitated by lead(II) acetate $Pb(CH_3COO)_2$. Al^{3+} ion is precipitated in the form of $Al(OH)_3$ when reducing the basicity of the solution and adding a strong base such as NaOH, thus Al^{3+} becomes more amphoteric than Cr^{3+} and Cr^{3+} will not precipitate with Al^{3+}.

$Cr(OH)_3$ has the largest solubility because it has a higher K_{sp} than $Al(OH)_3$. $Al(OH)_3$ has higher K_{sp} than $Fe(OH)_3$, therefore, Fe^{3+} is the first one to be precipitated, then Al^{3+}, and the last one is Cr^{3+}.

Iron (III) hydroxides are precipitated when $pH \leq 2$ which means it is precipitated in an acidic medium. Iron (III) hydroxides differ from aluminum hydroxide and chromium hydroxide as iron (III) hydroxides are not dissolved with an increase of the basicity of the solution. Therefore, it is a weak amphoteric.

NaOH cannot be used instead of NH_4OH because NH_4^+ could not be formed which is a common ion with NH_4Cl. The lack of NH_4^+ in the solution leads to losing control of the concentration of [OH] that is required for the precipitation.

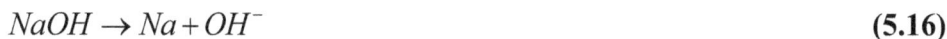

$$NH_4Cl \rightarrow NH_4^+ + Cl^- \tag{5.14}$$

$$NH_4OH \leftrightarrow NH_4^+ + OH^- \tag{5.15}$$

$$NaOH \rightarrow Na + OH^- \tag{5.16}$$

All interactions that occur are inorganic reactions. In order to detect Cr^{3+}, pentanol is used that will form an organic layer.

The Cations of group III are ($Cr^{3+}, Al^{3+}, Fe^{2+}$ and Fe^{3+}), add 0.2 mg of NH_4Cl and 4 drops of NH_4OH then heat it in water bath for 3 minutes then place it in a centrifuge for 5 minutes then filter the solution.

Filtrate is negligent as it contains other ions of group cations

precipitate contains $Cr(OH)_3$, $Fe(OH)_3$, $Al(OH)_3$, Add 20 drops of NaOH or NH_4OH and 10 drops of H_2O_2 then place it in a centrifuge for 5 minutes then filter the solution.

The precipitate contains $Fe(OH)_3$ add 5 drops of NaOH and add (2-3) drops of H_2O_2 then heat in water bath for 1 minute

Filtrate contains (Na_2CrO_2, Na_2AlO_2) and it is divided into three parts.

blue precipitate (Prussian blue) from $KFe[Fe(CN)_6]$

The solution is acidified by CH_3COOH and test by blue litmus paper. Lead(II) acetate $Pb(CH_3COO)_2$ (0.1 M) is added to the solution (2-3 drops). A yellow solid is precipitated $PbCrO_4$ as the indicating of the presence of Cr^{3+}

The solution is acidified by HCl and test by blue litmus paper. Then a mixture of NH_4Cl and NH_4OH is added. Heat for 1 minute. a white precipitate is formed as indication of $Al(OH)_3$

The solution is acidified by HNO_3 (7) drops then 8 drops of pentanol $C_5H_{11}OH$ is added then 7 drops of H_2O_2 is added. A blue organic layer is formaed from Cr^{3+}

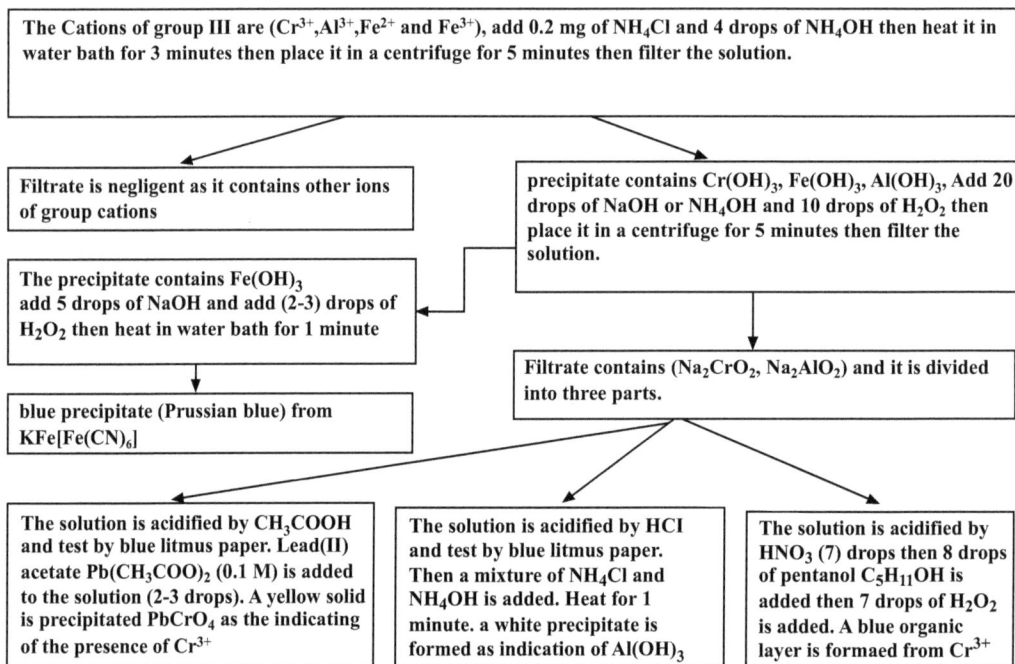

Flow chart (5.1). The analysis of cations group III.

CHAPTER 6

The Fourth Group

Abstract: This chapter discusses the method of detection and separation of the fourth group cations, precipitate format, the impact of pH on the procedure.

Keywords: Acidic medium, Black precipitate, Pink precipitate, Sulfides, White precipitate.

DETECTION AND SEPARATION OF THE FOURTH GROUP CATIONS

Group IV

This group contains Manganese Mn^{2+}, Cobalt Co^{2+}, Nickel Ni^{2+}, and Zinc Zn^{2+}. These ions are divalent. The group reagent is hydrogen sulfide in a weakly acidic medium. To control the acidic medium, the mixture of ammonium chloride NH_4Cl and ammonium hydroxide NH_4OH is used. Ammonium hydroxide is a weak base. However, the alkalinity of Ammonium hydroxide is high. Therefore, the alkalinity decreases with the addition of ammonium chloride, which leads to the following balance shift to the left [1, 5].

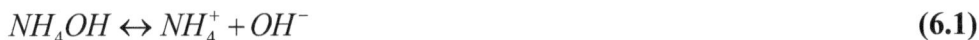

$$NH_4OH \leftrightarrow NH_4^+ + OH^- \tag{6.1}$$

The solution contains H_2S and ammonium chloride NH_4Cl and ammonium hydroxide NH_4OH. Therefore, another balance is as follows:

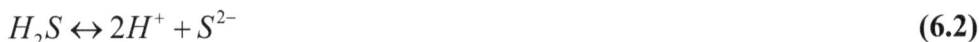

$$H_2S \leftrightarrow 2H^+ + S^{2-} \tag{6.2}$$

Since an appropriate amount of ammonium hydroxide is added, so the medium is basic $[OH] > 10^{-7}$. Therefore, the equilibrium in Equation 6.2 is shifted to the right so that the group reagent contains a good concentration of S^{2-} ion which is much larger than the concentration of S^{2-} of the group reagent for the second group because the medium is acidic in that group. Thus, the precipitated medium here contains S^{2-} ion and OH^- with a relatively higher concentration. Both of these ions are precipitated because sulfide and hydroxide are more insoluble minerals [1, 5].

The cations of this group are not precipitated when the cations of the second group are precipitated because these sulfides are dissolved in the acidic medium of HCl (0.3M) as observed previously. In other words, the concentration of S^{2-} ion is not sufficient for precipitation of the group IV ions in the conditions of the second group. The group IV ions are precipitated in the form of sulfides as follows:

$$Co(NO_3)_2 + H_2S_{(g)} \rightarrow CoS_{(s)} + 2HNO_3 \tag{6.3}$$

Black precipitate

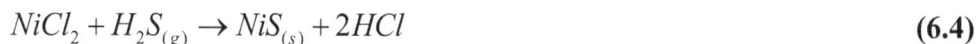

$$NiCl_2 + H_2S_{(g)} \rightarrow NiS_{(s)} + 2HCl \tag{6.4}$$

Black precipitate

Black precipitates of CoS and NiS are dissolved in hot nitric acid and Aqua regia, respectively. However, these precipitates are insoluble in hydrochloric acid.

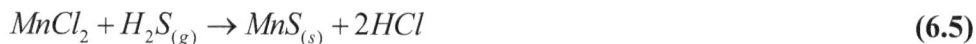

$$MnCl_2 + H_2S_{(g)} \rightarrow MnS_{(s)} + 2HCl \tag{6.5}$$

Pink precipitate

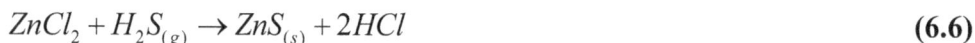

$$ZnCl_2 + H_2S_{(g)} \rightarrow ZnS_{(s)} + 2HCl \tag{6.6}$$

White precipitate

Both ZnS and MnS are dissolved in the mineral acids. Besides, zinc sulfide is dissolved in acetic acid.

Methodology

Group IV is precipitated and separated from the rest of the groups in the analysed mixture using the following steps (see Flow chart **6.1**):

1. Take filtered liquid produced by separating the ions of the third group, add 1-2 g of NH₄Cl, and then heat for 2-1 minutes on a water bath.
2. Add the ammonia solution to filtered liquid until it becomes alkaline (use red litmus paper, *i.e.*, change the color of the paper from red to blue). Then, add 2 drops of the thioacetamide solution and heat it for 15 minutes. The ions of this group will be precipitated as sulfides.

3. Separate the filtered liquid from the precipitate after placing a solution in the centrifuge for 5 minutes. Keep the filtered liquid to detect subsequent group ions if present.
4. Group IV ions are separated from each other and their ions are detected by following scheme (6.1).
5. The precipitate contains sulfide of insoluble cobalt, nickel, zinc, and magnesium. Add about 10 drops of HCl to the solution with continuous stirring and heat it on a water bath for 3-5 minutes, the following chlorides will be obtained:

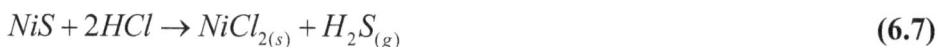

$$NiS + 2HCl \rightarrow NiCl_{2(s)} + H_2S_{(g)} \tag{6.7}$$

Precipitate chloride

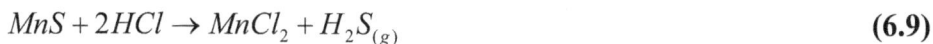

$$CoS + 2HCl \rightarrow CoCl_{2(s)} + H_2S_{(g)} \tag{6.8}$$

Precipitate chloride

$$MnS + 2HCl \rightarrow MnCl_2 + H_2S_{(g)} \tag{6.9}$$

Soluble chloride

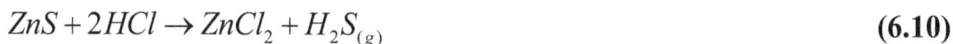

$$ZnS + 2HCl \rightarrow ZnCl_2 + H_2S_{(g)} \tag{6.10}$$

Soluble chloride

6. The filtered liquid ($ZnCl_2$, $MnCl_2$) is separated from the precipitate ($CoCl_2$, $NiCl_2$). Nine drops from the Aqua Regia ($1HNO_3$:$3HCl$) is added to the precipitate and heated for 1-2 minutes. Both precipitates will dissolve. The filtered liquid is divided into two parts:

The First Section: It detects the presence of Co^{2+} ion by adding a few drops of ammonium thiocyanate (NH_4SCN). A blue color solution appears indicating the presence of cobalt.

$$CoCl_2 + 2NH_4SCN \rightarrow (NH_4)_2[CoSCN_4] + NH_4Cl \tag{6.11}$$

The Second Section: The presence of Ni^{2+} ion is detected by the addition of (1spatula) $NH_4Cl_{(S)}$ and then the solution is made basic by NH_3 solution and tested with litmus paper. Then, 7 drops of the Dimethylglyoxime (DMG)

$C_4H_8N_2O_2$ is added. A red precipitate is formed indicating the presence of nickel, while cobalt ions form a brown dissolved complex with this organic compound.

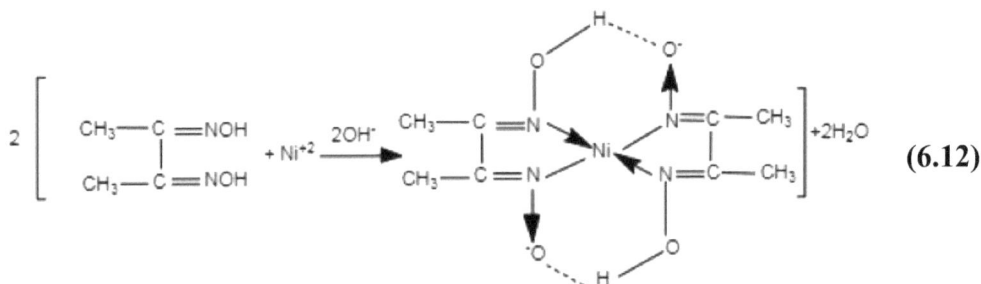

$$(6.12)$$

The filtered liquid contains chlorides ($MnCl_2$, $ZnCl_2$). The filtered liquid is boiled to remove the remaining H_2S. The removal of H_2S is tested by placing a wet paper of lead (II) acetate on the top of the test tube. Paper turns black indicating the presence of gas until the non-tarnishing paper indicates the removal of H_2S gas.

$$H_2S_{(g)} + Pb(CH_3COO)_2 \rightarrow PbS_{(s)} + 2CH_3COOH \qquad (6.13)$$

Black precipitate

Two to four drops of NaOH solution is added to the filtered liquid followed by 2 drops of H_2O_2. The solution is then heated for 3 minutes. The filtered liquid is divided into two halves:

First Part (Precipitate):   In order to test the presence of Zn^{2+} 20 drops of the solution of thioacetamide C_2H_5NS are added and then heated for 20 minutes giving a white precipitate indicating the presence of zinc.

$$ZnCl_2 + 2NaOH \rightarrow Zn(OH)_2 + 2NaCl \qquad (6.14)$$

Soluble

$$ZnOH_2 + H_2S_{(g)} \rightarrow ZnS_{(s)} + 2H_2O \qquad (6.15)$$

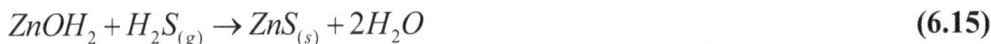

White precipitate

Second Part (Filtrate): It is used to detect the presence of manganese (II) ions Mn^{2+}. 250 mg of Pb_3O_4 red lead is added to the solution which yields dioxide in the presence of nitric acid. Then, it is heated for 15 minutes and then cooled and diluted with water. The solution turns red-violet (purple) indicating the presence of manganese.

$$5PbO_2 + 2Mn^{2+} + 4H^+ \rightarrow 2MnO_4^- + \rightarrow 5Pb^{2+} + 2H_2O \qquad (6.16)$$

ADDITIONAL INFORMATION:

1. Nitrate, chlorides, bromides, iodides, acetates, and sulfates of these ions are dissolved.
2. Sulfide and hydroxides of these ions are insoluble.
3. Zinc oxide (ZnO) is characterized by some amphoteric properties of oxides, while the other oxides do not dissolve in the basic solution, *i.e,* basic oxides.
4. The hydroxides of this group are not precipitated in a mixture of NH_4Cl + NH_4OH because the solubility constants are not very low compared to the solubility constituents of the hydroxides of the third group.

$Mn(OH)_2$	1.6×10^{-13}
$Co(OH)_2$	2×10^{-16}
$Ni(OH)_2$	3×10^{-19}
$Zn(OH)_2$	7×10^{-18}

5. Mn^{2+} is stable in acid solutions. Although it is oxidized by a strong oxidizing agent to MnO_2, which is insoluble if it is oxidized by chlorate, and it is oxidized to MnO_4 (violet) if it is oxidized by sodium bismuthate ($NaBiO_3$) or lead dioxide (PbO_2) in a strongly acidic medium.
6. Zinc ion does not normally give any color compounds either in solid-state or in solutions because it does not have any neighboring electrons. ZnS is the only metal sulfide that has a white color that is different from other mineral sulfides.
7. Part of the zinc ion is precipitated with the second group if the acidity of a solution is less than 0.3 normality during the precipitation of that group. Therefore, zinc sulfide dissolves in diluted HCl acid, which distinguishes it from nickel sulfide and cobalt sulfide.

8. Zinc oxide and hydroxide are amphoteric, and therefore dissolve in acids and strong bases.

9. Zinc hydroxide is also dissolved in ammonium hydroxide due to the formation of complex $Zn(NH_3)_4^{2+}$. Zinc ion forms many complexes. These complexes are disassociated by sulfide ion. H_2S precipitates Zn^{2+} and takes it from these complexities.

10. Co^{2+} ion is pink and is identified in aqueous solutions. It has the ability to form complex ions and has no amphoteric properties. The blue-colored $Co(SCN)_4^{2-}$ complex can be extracted by ethyl alcohol or ether.

11. Ni^{2+} ion and Co^{2+} ion are similar, *i.e.*, have colored water solutions. Ni^{2+} ion is a green-colored aqueous solution. Ni^{2+} ions can be dissolved in HCl and Aqua Regia and form complex with both NH_3 and SCN^-.

12. Sodium fluoride NaF is added to the mixture (Ni^{2+} and Co^{2+}) at the final stage of detection because NaF works to block Co^{2+} from Ni^{2+} by forming a stable complex $[NiF_6]^{4-}$, thereby leaving Co^{2+} which can be detected by adding NH_4SCN, where a blue complex $[Co(SCN)_4]^{2-}$ is extracted into an organic layer by alcohol.

13. NaF is a masking reagent. It is a blocking agent which is added to the interfering reaction. It has the ability to react with the interfering ions, forming stable complex compounds, leaving the other ions free to interfere with other ions. This reaction is known as a masking reaction.

14. Interfering Reaction: These are reactions that occur between some ions in the solution, competing with the ion to be detected in their interactions with the reagent (qualitative detector). These reactions occur in a shorter duration because the interfering ions are already transformed into stable complexes with the masking reagent, thus becoming inefficient when adding a qualitative reagent to identify the ion to be tested without the physical separation of the interfering ions.

15. The presence of iron ions interferes with the detection of Ni^{2+} ion and Co^{2+} ion. Therefore, the addition of sodium fluoride to a mixture of these ions leads to the conversion of ferric ions into $(FeF_6)^{3-}$ ions which do not react with the solutions that are used to detect both Ni^{2+} ion and Co^{2+} ion.

16. The blue complex ion $[Co(SCN)_4]^{2-}$ is unstable in aqueous solutions. Thus, a blue color ring will form at the meeting point with the added alcohol solution (NH_4SCN), where the color disappears upon shaking or stirring. Therefore it is preferable to add more than 10 drops of alcohol solution when performing the test.

The presence of cobalt ions in the detection of nickel ions is complicated by the formation of a dark complex with a dimethylglyoxime solution. Therefore, a reagent should be added to confirm the deposition of the nickel-like nickel-colored dimethylglyoxime in the case of nickel ions.

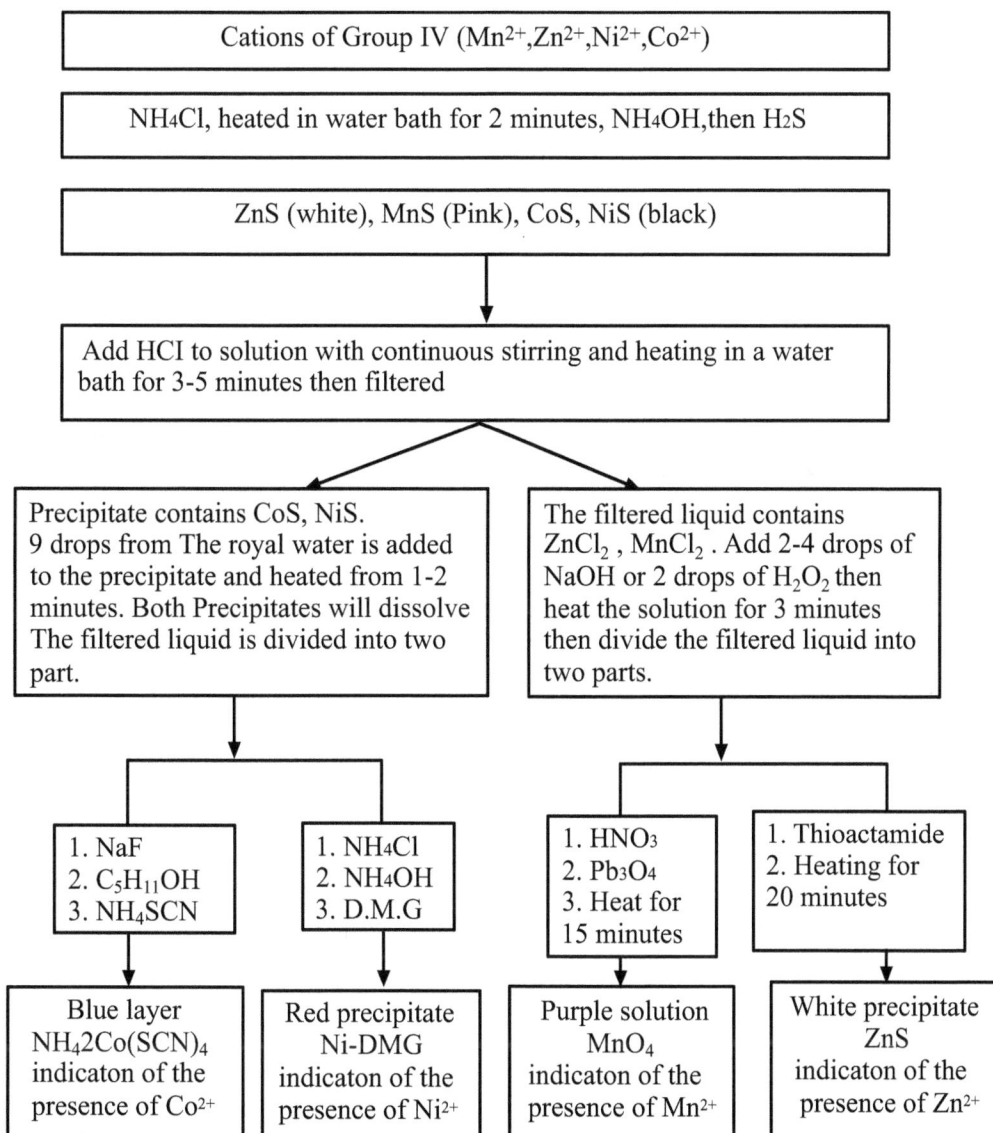

Cations of Group IV (Mn^{2+}, Zn^{2+}, Ni^{2+}, Co^{2+})

NH_4Cl, heated in water bath for 2 minutes, NH_4OH, then H_2S

ZnS (white), MnS (Pink), CoS, NiS (black)

Add HCl to solution with continuous stirring and heating in a water bath for 3-5 minutes then filtered

Precipitate contains CoS, NiS. 9 drops from The royal water is added to the precipitate and heated from 1-2 minutes. Both Precipitates will dissolve The filtered liquid is divided into two part.	The filtered liquid contains $ZnCl_2$, $MnCl_2$. Add 2-4 drops of NaOH or 2 drops of H_2O_2 then heat the solution for 3 minutes then divide the filtered liquid into two parts.

1. NaF 2. $C_5H_{11}OH$ 3. NH_4SCN	1. NH_4Cl 2. NH_4OH 3. D.M.G	1. HNO_3 2. Pb_3O_4 3. Heat for 15 minutes	1. Thioactamide 2. Heating for 20 minutes

Blue layer $NH_42Co(SCN)_4$ indicaton of the presence of Co^{2+}	Red precipitate Ni-DMG indicaton of the presence of Ni^{2+}	Purple solution MnO_4 indicaton of the presence of Mn^{2+}	White precipitate ZnS indicaton of the presence of Zn^{2+}

Flow chart (6.1). The analysis for cations group IV.

CHAPTER 7

The Fifth Group

Abstract: This chapter elaborates on the the detection and separation of the fifth group cations. It explains step-by-step practical directions and a number of rules that must be followed for successful detection and separation of the cations.

Keywords: Ammonium chloride, Ammonium hydroxide, Carbonates, White precipitate.

DETECTION AND SEPARATION OF THE FIFTH GROUP CATIONS

Group V

This group includes calcium ions Ca^{2+}, strontium ions Sr^{2+} and barium ions Ba^{2+}, which belong to alkaline earth metals in the periodic table. These metals produce soluble chlorides, sulphides and hydroxides in conditions that allow precipitation of groups I, II and III. However, these ions precipitate in the form of carbonates using ammonium carbonate (buffer solution) from ammonium hydroxide and ammonium chloride [1, 5].

The purpose of using ammonium chloride in the mixture of [NH₄Cl + NH₄OH + (NH₄)₂CO₃] is to precipitate ions of this group and control the number of carbonate ions that are precipitated so that the magnesium ions are not precipitated with this group, neither in the form of carbonate nor in the form of hydroxide. The presence of ammonium ions in a large quantity reduces the concentration of carbonates precipitated by moving the equilibrium to the right. Thus, it possible to get rid of the carbonate where the bicarbonate gets dissociated when the solution is heated. The bicarbonates are converted into carbonates by neutralizing the ammonia solution [5].

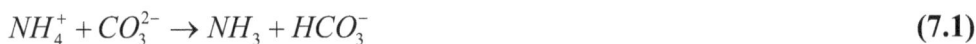

$$NH_4^+ + CO_3^{2-} \rightarrow NH_3 + HCO_3^- \tag{7.1}$$

The presence of ammonium in a large amount reduces the concentration of hydroxyl ions that are produced when the ammonium hydroxide is ionised.

Huda S. Alhasan & Nadiyah Alahmadi

$$NH_4OH \leftrightarrow NH_4^+ + HO^- \qquad\qquad (7.2)$$

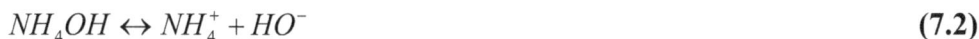

But the concentration of ammonium ion NH_4^+ should not be too large that may have an impact on the equilibrium.

If the equilibrium moves to the right, this could lead to reducing the concentration of carbonates to the level that is not sufficient to precipitate the ions of group V. The presence of ammonium hydroxide prevents the previous behaviour, as a consequence of hydroxides ions that are produced from the dissociation of ammonium hydroxide combined with the increase of ammonium ions, that leads to the equilibrium move to the left.

In other words, the buffer solution ($NH_4OH + NH_4Cl$) provides an appropriate concentration of carbonate ions that is sufficient to precipitate the ions of group V but is insufficient to precipitate magnesium carbonate.

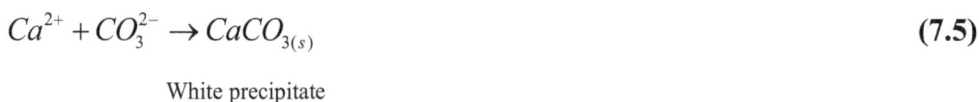

$$Ba^{2+} + CO_3^{2-} \rightarrow BaCO_{3(s)} \qquad\qquad (7.3)$$
White precipitate

$$Sr^{2+} + CO_3^{2-} \rightarrow SrCO_{3(s)} \qquad\qquad (7.4)$$
White precipitate

$$Ca^{2+} + CO_3^{2-} \rightarrow CaCO_{3(s)} \qquad\qquad (7.5)$$
White precipitate

Warning: Do not direct the test tube while heating up toward you or toward your colleagues, and heat the centre of the tube with the continuous movement to distribute the heating on all parts of the liquid.

The Method of Work

Group V ions are precipitated and separated from the group VI ions by the following steps (See [9]):

1. Heating the filtered liquid produced by filtering group IV at the boiling point in order to remove ammonium salts and then diluting the solution by adding a small amount of distilled water. Following this, the medium of the solution is

changed to make it basic by adding several drops of NH₄OH followed by the addition of a few drops of NH₄Cl and about 4-5 drops of NH₄CO₃. The mixture is heated in a water bath for 2 minutes.

2. Filtering the mixture and separating the precipitates that contain the ions of group V in the form of carbonates. The filtered liquid is retained that contains the ions of group V.

The cations of group V are separated from each other and then detected. (see Flow chart **7.1**).

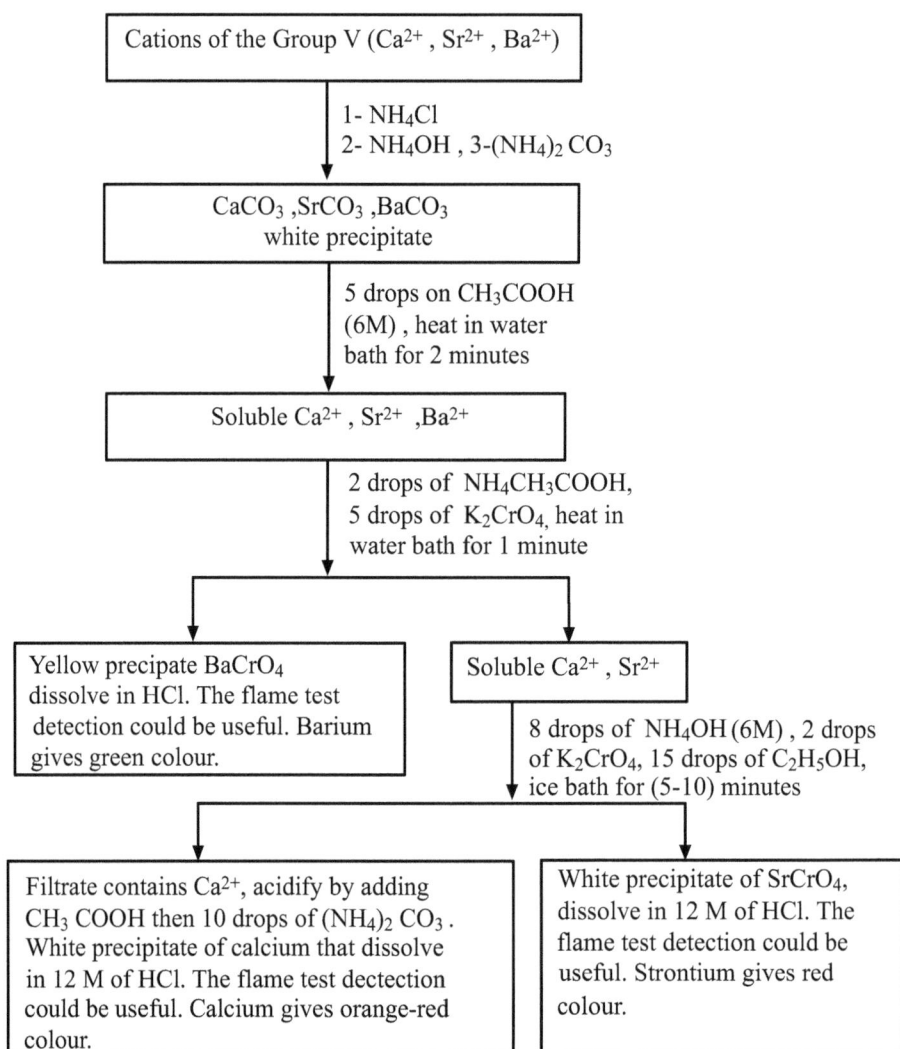

Flow chart 7.1. The analysis of cations group V.

ADDITIONAL INFORMATION:

1. The ions of group V belong to the alkaline earth metals group. Radium, which is the last element in this group has similar chemical properties to barium. However, it is not included in this group because of its scarcity and radiation properties.
2. Magnesium is an alkaline earth metal. Its hydroxide is less basic and its carbonates are more soluble than the carbonate of alkaline earth metals.
3. The ions of group V have eight electrons in their last orbit so their ions are not coloured. Since alkali ions are not coloured, their compounds are also not coloured. They become coloured if they contain a coloured ion such as a chromate ion.
4. Hydroxides of alkaline earth metals are medium or well-soluble in water; therefore, they do not precipitate when one of the strong bases is added to the weak base of solutions of alkaline earth salts but they are precipitated if the solutions are concentrated as well as the strong basic.
5. Since alkaline reagents absorb CO_2 from the atmosphere, they give carbonate precipitates that are added to alkaline earth metal salts.
6. Carbonates of alkaline earth metal are formed when dissolved carbonate is added to an alkaline or mild solution that contains ions of an alkaline earth metal.
7. These carbonates dissolve in strong or weak acid solutions such as acetic acid and carbonic acid.

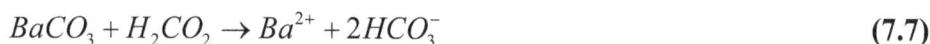

$$CaCO_3 + 2HC_2H_3O_2 \rightarrow Ca^{2+} + 2C_2H_3O_2^- + H_2O + CO_2 \qquad \textbf{(7.6)}$$

$$BaCO_3 + H_2CO_2 \rightarrow Ba^{2+} + 2HCO_3^- \qquad \textbf{(7.7)}$$

For this reason, precipitation should be done in a mild or alkaline medium.

1. The increase in the concentration of ammonium ions should be eliminated before precipitation by heating because the increased concentration of ammonium ions reduces the concentration of carbonate ions by the following reaction:

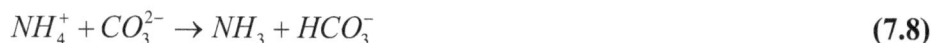

$$NH_4^+ + CO_3^{2-} \rightarrow NH_3 + HCO_3^- \qquad \textbf{(7.8)}$$

2. Barium and strontium precipitate immediately when H_2SO_4 acid or sulphate salt solution is dissolved into a solution of one ion, even if the solution is diluted. Calcium sulphate $CaSO_4$ is precipitated only in the concentrated or medium concentration of calcium solution.

3. Barium easily forms yellow precipitates with potassium chromate. Strontium precipitates only in a concentrated solution or in the presence of ethyl alcohol, which reduces the electrostatic insulation of the medium and calcium.

4. We acquire the best results in detecting group V ions by the flame test when using chloride salts of the ions because they are more volatile than other salts.

5. The best way to conduct a flame test detection is to dip a platinum wire into an ion chloride solution to test the ions and then expose it to the flame of the Bunsen burner. In case if platinum wire is not available, a piece of filter paper could be useful. Take filter paper in the shape of a pen can be dipped into the solution and then exposed to the flame. However, we should not rely on the flame test detection alone without using the method of analysis of this group. The flame test detection is to ensure the existence of ions.

The reason for the appearance of colour of the ions when exposed to flame is due to the fact that the heat from the Bunsen burner excites an electron so it moves from one energy orbit (ground state) to another energy orbit (excited state) that is away from the nucleus. This is an unstable state; therefore, an electron moves back to its ground state and releases a certain amount of energy in the form of light (emission).

CHAPTER 8

The Sixth Group

Abstract: This chapter focuses on the sixth group cations (detection and separation), and elaborates on the method and step-by-step practical directions for the flame test.

Keywords: Alkali ions, Ammonium chloride, Flame test, Group reagent.

DETECTION AND SEPARATION OF THE SIXTH GROUP CATIONS

Group VI

This group contains magnesium ions Mg^{2+}, sodium Na^+, potassium K^+ and ammonium NH_4^+. These ions are not precipitated with any of the previous *via* groups. Magnesium ions may be partially precipitated in the form of a hydroxide or a carbonate with the fourth and fifth groups if the amount of ammonium chloride is not added sufficiently during precipitation [5, 9].

There is no general group regent for group VI. Therefore, all common sodium, potassium and ammonium salts are soluble in water. There are some complexes that have intermediate solubility, as the result of the presence of alkali ions in their composition. These complexes are used in the detection of ions. We will talk about each while talking about the analytical properties of each ion.

In case of magnesium ions, its hydroxide, carbonate, phosphates, and oxalate are insoluble in water. Therefore, this method is used for the detection of Mg^{2+} ions [1, 5].

THE ANALYTICAL PROPERTIES OF THE IONS OF THIS GROUP

Mg^{2+} have ions similar to the group of alkaline earth minerals, while sodium and potassium follow the group of alkali minerals. Since NH_4^+ is similar to the K^+ ion, so it is classified in this group [9].

The metals of this group are active metals since they easily lose the electrons from the outer surface. Most of their ionic compounds are solid. Alkali ions seldom form complex compounds because of their large size and a small charge. The ions are colourless because there are no unpaired electrons in the last layer.

Huda S. Alhasan & Nadiyah Alahmadi

Caution: Since sulfuric acid and potassium dichromate are very acidic, their mixture has similar properties. We must avoid using this solution, and it could be replaced by known commercial cleaning solutions.

THE WORK METHOD

Detection of Mg^{2+}

Firstly, it is important to make sure that all the former groups' ions especially barium and strontium ions, were isolated during the analysis of these groups by adding a mixture of 0.5 mL of ammonium sulphate $(NH_4)_2SO_4$ and 0.5 mL of ammonium oxalate $C_2H_8N_2O_4$ to the filtrate of the fifth group [1, 9].

Following this, the mixture is heated and left aside for a few minutes and then the solution is filtered whenever the precipitate appears. The supernatant contains the ions of the sixth group. The filtrates are kept and and the precipitates are discarded.

NH_4OH is added to the filtrate until the medium becomes basic, followed by the addition of five more drops of NH_4OH. After this, 1 mL of Na_2HPO_4 is added and the solution is stirred well with a stirring rod to scratch the walls with the test tube. After a few minutes of wait, if a white crystalline solid appears, this is taken as evidence of the presence of Mg^{2+} (see Flow chart **8.1**).

Detection of NH_4^+

3 mL of the original solution is taken and placed in an evaporating dish. After this, 3 mL of concentrated NaOH or KOH is added and mixed well with a stirring rod and covered with a watch glass containing a piece of moist red litmus paper inside. It is heated gently until the vapors start to release. If the colour of the paper uniformLy turns into blue, this indicates the presence of ammonium in the solution. (see Flow chart **8.2**)

Detection of Na^+

In 3 mL of the original solution in a test tube, a concentrated solution of KOH is added until it becomes alkaline. Following this, about 3 mL of a concentrated solution of K_2CO_3 is added and boiled for a short period, and then cooled down and filtered to make sure that the precipitation process is complete. Following this, the supernatant is placed in the crucible, heated over a gentle fire to dryness. The remaining salt in the crucible is dissolved with 1.5 mL of water. Sodium is detected by either by adding uranyl acetate or magnesium acetate or antimony potassium tartrate. First a yellowish-green precipitate is formed and then a white precipitate. In the flame test, an intense yellow bulky flame appears if Na^+ is present.

Detection of K^+

In 3 mL of the original solution into a test tube, a concentrated solution of NaOH is added until it becomes alkaline.

After this 3 mL of concentrated solution Na_2CO_3 is added and the mixture is heated. The mixture in then cooled down and filtered to make sure that the precipitation process is complete. The supernatant is placed in the crucible, heated over a gentle fire to dryness. The remaining salt in the crucible is dissolved with 1 mL of water and divided into two equal portions (see Flow chart **8.2**).

The First Portion: Several drops of acetic acid are added until the solution is slightly acidic followed by the addition of 1 mL of sodium cobalt nitrite $Na_3Co(NO_2)_6$. The appearance of a yellow precipitate indicates the presence of K^+.

The Second Portion: A flame test is carried out, where a violet flame appears if K^+ is present.

ADDITIONAL INFORMATION:

1. The detection of sodium is a very sensitive process since the residues of sodium ions that are found in most solutions and even in glass bottles could make the solutions attain a colour. Therefore, it is necessary to compare the detection of the unknown solution with a solution that contains sodium ions. If intense yellow colour appears for five seconds or more, that is the evidence of the presence of sodium ions, but if the colour of yellow is not dense and remains for a short period (less than five seconds), sodium ions are not present [1, 9].
2. Potassium compounds are more volatile than sodium compounds. Therefore, the potassium compounds take less time to produce a colour in flame than the sodium compounds. Violet colour produced by the potassium flame is changed by the yellow colour of the sodium flame.
   It is, therefore, possible to observe the colour of the flame through a cobalt glass as it absorbs the yellow colour and releases the red-violet colour. The flame

from the unknown solution should be compared with a known solution for the same reason mentioned in case of the detection of sodium ions.

3. The presence of ammonium ions with hydroxide ions leads to the release of ammonia gas. However, the high concentrations of both hydroxide ions and ammonium ions increase the concentration of ammonia gas that is released. Ammonia gas is less soluble in hot solutions than cold solutions and turns the moist red litmus paper into blue.

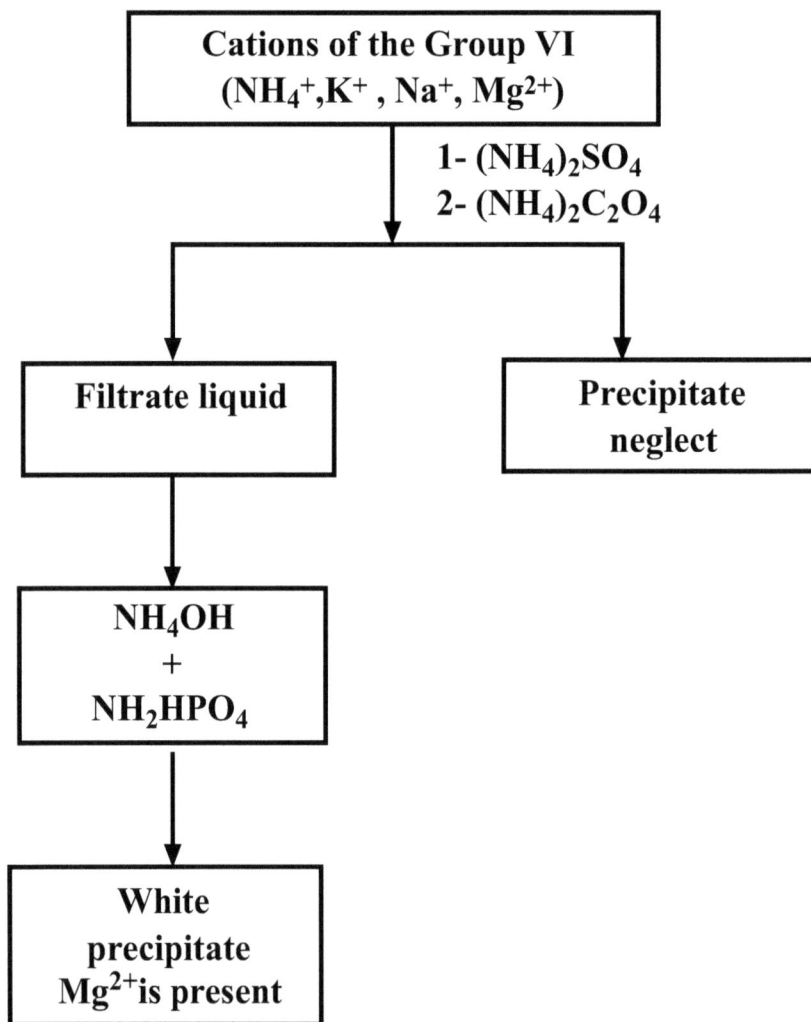

Cations of the Group VI
(NH_4^+, K^+, Na^+, Mg^{2+})

1- $(NH_4)_2SO_4$
2- $(NH_4)_2C_2O_4$

Filtrate liquid   **Precipitate neglect**

NH_4OH
+
NH_2HPO_4

White
precipitate
$Mg^{2+}$is present

Flow chart (8.1). The analysis of cations group VI (detection of Mg^{2+}).

```
                              ┌──────────────────┐
                              │ Original solution │
                              └──────────────────┘
```

- Original solution
  - Concentrated Na$_2$CO$_3$ + Concentrated NaOH
    - acetic acid and Sodium cobaltinitrite — A yellow precipitate K$^+$ is present
    - flame test, violet flame — K$^+$ is present
    - Precipitate neglect
  - Concentrated K$_2$CO$_3$ + Concentrated NaOH
    - Filtrate liquid — Heat — Dissolve the remaining salt with water
      - Antimony potassium tartrate. A white precipitate Na$^+$ is present
      - flame test an intense yellow bulky flame Na$^+$ is present
      - uranyl zinc acetate or uranyl magnesium acetate A yellowish green precipitate Na$^+$ is present
    - Precipitate neglect
  - Concentrated KOH — Heat
    - Red litmus paper
      - The colour of the paper turn to uniformly blue, NH$_4$$^+$ is present

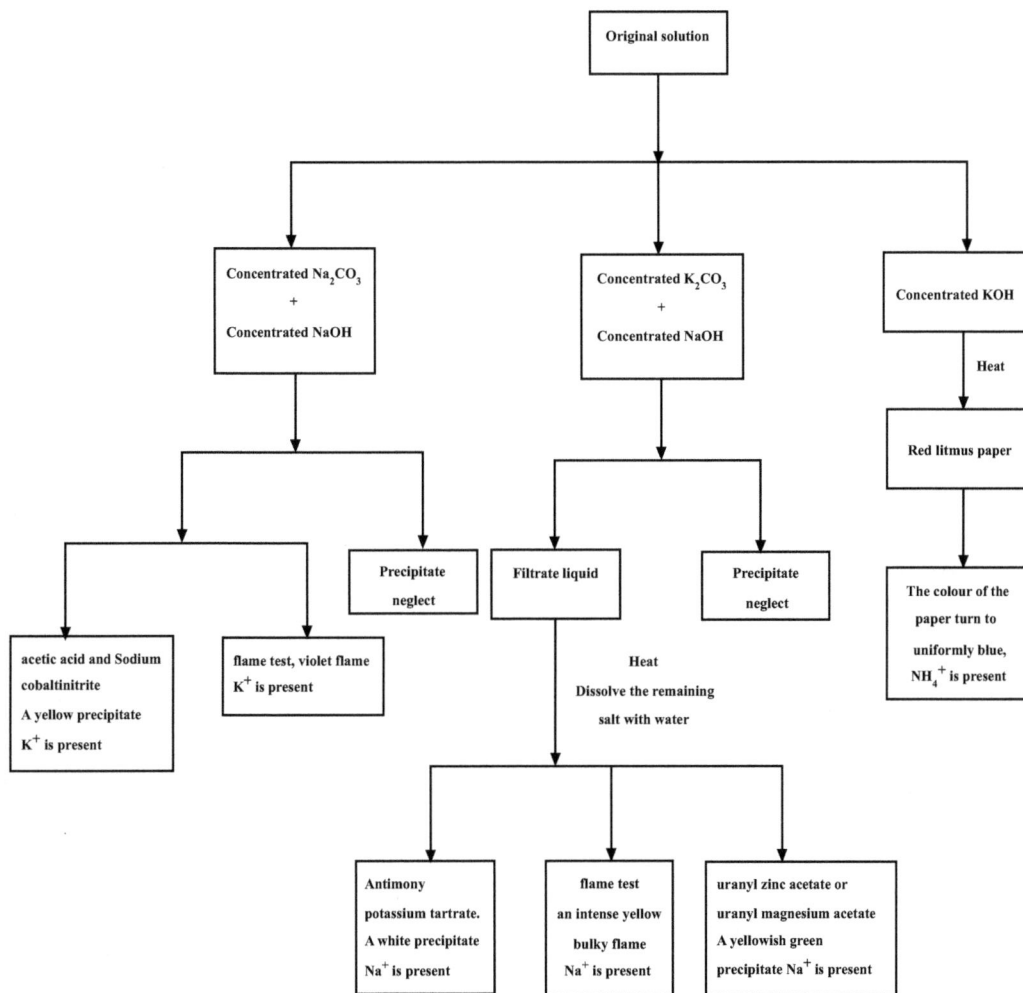

Flow chart (8.2). The analysis of cations group VI (detection of K$^+$, Na$^+$, NH$_4$$^+$).

CHAPTER 9

Summary of Groups Analysis with General Questions

SUMMARY ANALYSIS OF ALL GROUPS CATIONS (UNKNOWN SOLUTION)

Precipitates contain the cations of the first group PbCl₂, Hg₂Cl₂, AgCl White precipitate	Add 2M of HCl acid to an unknown solution, then cool the solution and filter it; make sure to acidify the filtered liquid.
Precipitate shows the existence of the cation of the second group.HgS (black), CuS(black), Bi₂S₃(brown), SnS(black) , SnS (yellow), CdS (yellow), Sb₂S₃ (orange), PbS(black)	Pass H₂S₍g₎ to the acidic solution using thioacetamide. The first group can be precipitated to form sulphur in the form of yellow precipitation of H₂S
Precipitate shows the existence of the cation of the third group. Al(OH)₃ (white), Cr(OH)₃ (green) , Fe(OH)₃ (brown)	Heat the filtered liquid to remove H₂S, then add cons of HNO₃ and heat it again to convert Fe²⁺ into Fe³⁺, then add 2M of NH₄Cl, 2 M of NHOH to make the solution basic.
Precipitate shows the existence of the cation of the Fourth group. CoS (black), NiS(black), MnS(pink), ZnS(white).	Add 2M of NH₄Cl, 2 M of NHOH to the filtered liquid then pass H₂S₍g₎
Precipitate shows the existence of the cation of the fifth group. CaCO₃ (white) , SrCO₃ (white) , BaCO₃ (white)	Heat the filtered liquid to remove H₂S, then add 2 M of (NH₄)₂CO₃, 2M of NH₄Cl, 2 M of NHOH
Sixth group: Mg²⁺ , Na⁺ , K⁺	The filtered liquid of the fifth group contains the cations of the sixth group.

* If the pH of the solution is too low, cadmium and zinc can be precipitated.
*The main reason for the formation of sulphur is iron (III) ions triangular, chromate ions or sulfite ions.

SUMMARY OF THE PROPERTIES OF CATIONS GROUPS (GROUP I - GROUP VI)

Groups	Ions	Groups Regent	Common Characteristic
First	Silver Ag⁺, mercuros (I) Hg₂²⁺, lead Pb²⁺	HCl	Their chlorides do not dissolve in hydrochloric acid
Second	Mercuric(II) Hg²⁺, copper Cu²⁺, bismuth Bi³⁺, cadmium Cd²⁺, Arsenic As³⁺, Tin (Sn⁴⁺, Sn²⁺)	H₂S in acidic medium, *i.e.*, HCl	Their sulfides do not dissolve in HCl acid, but the last three sulfide elements dissolve in (NH₄)₂ S

(Table cont).....

Third	Iron Fe^{3+}, Aluminium Al^{3+}, Chrome Cr^{3+}	NH_4Cl and NH_4OH	Precipitate in the form of hydroxides
Fourth	Zinc Zn^{2+}, Magnesim Mn^{2+}, Nickle Ni^{2+}, Cobalt Co^{2+}	NH_4Cl, NH_4OH and H_2S	Precipitate in the form of sulphides in the basic medium.
Fifth	Calcium Ca^{2+}, Strontium Sr^{2+}, Barium Ba^{2+}	NH_4Cl, NH_4OH and $(NH_4)_2CO_3$ as well as the flame test.	Precipitate in the form of carbonate dissolved in mineral acids.
Sixth	Ammonium NH_4^+, sodium Na^+, potassium K^+, magnesium Mg^{2+}	They did not have a specific group regent but the flame test can be characteristic of some of them.	They do not form precipitate expect magnesium.

GENERAL QUESTIONS

1. How to separate the Ag^+ silver ion from lead ion into the mixture:

    a) An unknown solid that is likely to be $AgNO_3$ or Hg_2Cl_2 or $pb(CH_3COO)_2$. Will this be unidentified if it gives purple-red precipitate when adding K_2CrO_4 to this solution?

    b) Precipitate also turns white when adding HCl.

2. Are silver halides affected by light? Describe the results.
3. Why the acidity should not be greater than (0.3 N) when cation group II is precipitated? Why should not it be much lower than 0.3 N?
4. Solution contains ions: Cu^{2+}, pb^{2+}, Bi^{3+}, Hg^{2+}

    a) Which ion will precipitate when a few of NH_4OH ions are added to the previous solution so that it becomes weak alkaline?

    b) What precipitates when HCl is added to this solution?

    c) What is precipitated when the NH_4OH is added to the original solution?

5. How to separate the ions from each other?
6. Which is the reagent that dissolves CuS and does not dissolve HgS?
7. Which is the reagent that dissolves $Cu(OH)_2$ and does not dissolve $Bi(OH)_3$?
8. Write the chemical equations that represent:

    a) The solubility of HgS in Aqua Regia.

    b) Effect of concentrated NH_4OH on copper ion?

9. Using ammonium sulphide solution,

    a) How to separate the ions of the cation group II?

10. What are the ions that have an amphoteric property? Explain this with the chemical equations.

11. Why does $Al(OH)_3$ dissolve in an excess amount of NaOH and NH_4OH?

12. Which reagent is to be used to separate the following pairs?

    a) Cr^{3+}, Fe^{3+}

    b) Al^{3+}, Fe^{3+}

13. Which cations of group IV

    a) form a complex with ammonia?

    b) form amphoteric hydroxides?

14. If adding an excess amount of ammonium hydroxide to an unknown solution that may contain Co^{2+}, Ni^{2+}, Mn^{2+}, no precipitate is formed. What is your conclusion?

15. The acidic solution contains the following ions:
(Ni^{2+}, Al^{3+}, Sb^{3+}, Ag^+)

    a) 5 drops of this solution are added in a test tube and then about 2 ml of water is added, and white precipitates appear. Explain your result.

    b) What happens when you add a little of NH_4OH to this solution?

    c) What happens if an excess amount of NH_4OH is then added?

    d) What will be precipitated in the original solution if the pH is 0.3 and H_2S is passed through the solution?

    e) How to separate these five ions from each other?

    f) What will be precipitated in the original solution if it is alkaline and H_2S free?

16. Cations of group II and group IV are precipitated in the form of sulfides, discuss these terms, indicating the fundamental difference between them.

17. What ions are precipitated of cations of group V? What is the purpose of adding NH_4Cl and NH_4OH?

18. Why sodium carbonate should not be used in the precipitation of cations of group V?

19. Why we should not rely on the flame test detection only in the investigation of the existence of cation group IV?

20. Why extensive heating must be avoided during the precipitation of the cation group V?

21. In an acidic solution containing Hg_2^{2+}, Ba^{2+}, Ni^{2+}

   a) What is the colour of the solution?

   b) What is precipitated by adding an excess amount of NH_4OH? What is the colour of the precipitate?

   c) How do these ions are separated and how to make sure they are separated?

22. How to distinguish between calcium, strontium and barium salts?

23. How does a solid mixture of NH_4Cl and KCl be separated without using any reagent?

24. Why should NH_4^+ be detected in the original solution and not in the filtered liquid of the cation group V? What is the purpose of adding ammonium sulphate and ammonium oxalate to the filtered liquid of the cation group V prior to detection of magnesium?

APPENDIX

APPENDIX

Table 1. The solubility constant for compounds at 25°C.

Compounds	Formula	K_{sp}
Barium carbonate	$BaCO_3$	2.58×10^{-9}
Barium sulphate	$BaSO_4$	1.08×10^{-10}
Cadmium carbonate	$CdCO_3$	1.0×10^{-12}
Cadmium hydroxide	$Cd(OH)_2$	7.2×10^{-15}
Calcium carbonate	$CaCO_3$	3.36×10^{-9}
Calcium hydroxide	$Ca(OH)_2$	5.02×10^{-6}
Calcium phosphate	$Ca_3(PO_4)_2$	2.07×10^{-33}
Calcium sulphate	$CaSO_4$	4.93×10^{-5}
Cobalt(II) hydroxide	$Co(OH)_2$	5.92×10^{-15}
Copper(I) chloride	$CuCl$	1.72×10^{-7}
Copper(I) thiocyanate	$CuSCN$	1.77×10^{-13}
Lead(II) carbonate	$PbCO_3$	7.40×10^{-14}
Nickel(II) carbonate	$NiCO_3$	1.42×10^{-7}
Nickel(II) hydroxide	$Ni(OH)_2$	5.48×10^{-16}
Silver(I) carbonate	Ag_2CO_3	8.46×10^{-12}
Strontium carbonate	$SrCO_3$	5.60×10^{-10}
Zinc hydroxide	$Zn(OH)_2$	3×10^{-17}

Table 2. List of some acids.

Name	Formula
Acetylsalicylic acid	$CH_3CO_2\text{-}C_6H_4\text{-}CO\underline{O}\underline{H}$
Adipic acid	$\underline{H}OOC\text{-}(CH_2)_4\text{-}CO\underline{O}\underline{H}$
Benzoic acid	$C_6H_5\text{-}CO\underline{O}\underline{H}$
Chlorobenzoic acid	$ClC_6H_4\text{-}CO\underline{O}\underline{H}$
Malic acid	$\underline{H}OOC\text{-}CHOH\text{-}CH_2\text{-}CO\underline{O}\underline{H}$

(Table 2) cont.....

Malonic acid	$\underline{H}OOC\text{-}CH_2\text{-}COO\underline{H}$
Oxalic acid dehydrate	$\underline{H}OOC\text{-}COO\underline{H}. 2H_2O$
Phenylacetic acid	$C_6H_5\text{-}CH_2\text{-}COO\underline{H}$
Salicylic acid	$\underline{H}O\text{-}C_6H_4\text{-}COO\underline{H}$
Succinic acid	$\underline{H}OOC\text{-}(CH_2)_2\text{-}COO\underline{H}$
Tartaric acid	$\underline{H}OOC\text{-}CHOH\text{-}CHOH\text{-}COO\underline{H}$
Trichloroacetic acid	$Cl_3C\text{-}COO\underline{H}$

Note: *The acidic hydrogen atoms are underlined.*

Table 3. List of acids with their information.

Acid	Sp.gr	Percentage	M.Wt
HCl	1.18	36 %	36.45
HNO_3	1.42	70 %	63
H_2SO_4	1.835	98 %	98.06
H_3PO_4	1.75	85 %	97.97
$HClO_4$	1.54	61 %	100.45
HI	1.70	75 %	127.9
HBr	1.49	48 %	80.9
HF	1.125	98 %	20
CH_3COOH	1.05	99.5 %	60

Table 4. List of strong acids and weak acids.

Strong Acids	Weak Acids
Hydrochloric acid, HCl	Acetic acid, CH_3COOH
Hydrobromic acid, HBr	Hydrocyanic acid, HCN
Hydriodic acid, HI	Hydrofluoric acid, HF
Nitric acid, HNO_3	Nitrous acid, HNO_2
Sulfuric acid, H_2SO_4	Sulphurous acid, H_2SO_3
Perchloric acid, HClO	Hypochlorous acid, HOCl
Periodic acid, HIO_4	Phosphoric acid, H_3PO_4

Table 5. List of strong Bases and list of weak Bases.

Strong Bases	Weak Bases
Sodium hydroxide NaOH	Ammonia NH_3
Potassium hydroxide KOH	Sodium carbonate Na_2CO_3
Calcium hydroxide Ca(OH)$_2$	Potassium carbonate K_2CO_3
Barium hydroxide Ba(OH)$_2$	Aniline $C_6H_5NH_2$
Sodium phosphate Na$_2$PO$_4$	Trimethylamine $(CH_3)_3N$

Table 6. Conjugate Acid-Base Pairs.

Acid	Conjugate Base
Hydrochloric acid HCl	Chloride ion Cl^-
Sulfuric acid H_2SO_4	Hydrogen sulphate ion HSO_4^-
Hydronium ion, H_3O^+	Water H_2O
Hydrogen sulphate ion HSO_4^-	sulphate ion SO_4^{2-}
hypochlorous acid HOCl	Hypochlorite ion ClO^-
Dihydrogen phosphate ion $H_2PO_4^-$	Monohydrogen phosphate ion HPO_4^{2-}
Ammonium ion NH_4^+	Ammonia NH_3
Hydrogen carbonate ion HCO_3^-	Carbonate ion CO_3^{2-}
Water H_2O	Hydroxide ion OH^-
Conjugate acid	**Base**

Table 7. The concentration of strong acids and bases and the value of Kw.

Solution	$[H_3O^+]$M	$[HO^-]$M	$K_w = [H_3O^+][HO^-]$
Pure water	1.0×10^{-7}	1.0×10^{-7}	1.0×10^{-14}
0.10 M strong acid	1.0×10^{-1}	1.0×10^{-13}	1.0×10^{-14}
0.010 M strong acid	1.0×10^{-2}	1.0×10^{-12}	1.0×10^{-14}
0.10 M strong base	1.0×10^{-13}	1.0×10^{-1}	1.0×10^{-14}

(Table 7) cont.....

0.010 M strong base	1.0×10^{-12}	1.0×10^{-2}	1.0×10^{-14}
0.025 M strong acid	2.5×10^{-2}	4.0×10^{-13}	1.0×10^{-14}
0.025 M strong base	4.0×10^{-13}	2.5×10^{-2}	1.0×10^{-14}

Table 8. pH values at specified [H$_3$O$^+$].

Solution	[H$_3$O$^+$]M	pH
1.00 L of H$_2$O	1.0×10^{-7}	7.00
0.100 mol HCL in 1.00 L of H$_2$O	1.0×10^{-1}	1.00
0.0100 mol HCL in 1.00 L of H$_2$O	1.0×10^{-2}	2.00
0.100 mol NaCl in 1.00 L of H$_2$O	1.0×10^{-7}	7.00
0.0100 mol NaOH in 1.00 L of H$_2$O	1.0×10^{-12}	12.00
0.100 mol NaOH in 1.00 L of H$_2$O	2.5×10^{-13}	13.00

Table 9. Transition ranges of some indicators.

Indicator Name	Acid Colour	Transition Rang (pH)	Base Colour
Thymol blue	Red	1.2-2.8	Yellow
Methyl orange	Red	3.1-4.4	Orange
Litmus	Red	5.0-8.0	Blue
Bromothymol blue	Yellow	6.0-7.6	Blue
Thymol blue	Yellow	8.0-9.6	Blue
Phenolphthalein	Colourles	8.0-9.6	Red
Alizarin yellow	Yellow	10.1-12.0	Red

Table 10. Relative strengths of acids and Bases.

Acid	Formula	K_a of Acid	Conjugate Base	Formula
Hydronium ion	H_3O^+	5.53×10^1	Water	H_2O
Hydrogen sulphate ion	HSO_4^-	1.23×10^{-2}	Sulphate ion	SO_4^{2-}
Formic acid	$HCOOH$	1.82×10^{-4}	Formate ion	$HCOO^-$
Benzoic acid	C_6H_5COOH	6.46×10^{-5}	Benzoate ion	$C_6H_5COO^-$
Acetic acid	CH_3COOH	1.75×10^{-5}	Acetate ion	CH_3COO-
Carbonic acid	H_2CO_3	4.30×10^{-7}	Hydrogen carbonate ion	HCO_3^-
Dihydrogen phosphate ion	$H_2PO_4^-$	6.31×10^{-8}	Monohydrogen phosphate ion	HPO_4^{2-}
Hypochlorous acid	$HOCl$	2.95×10^{-9}	Hypochlorite ion	ClO^-
Ammonium ion	NH_4^+	5.57×10^{-10}	Ammonia	NH_3
Hydrogen carbonate ion	HCO_3^-	4.68×10^{-11}	Carbonate ion	CO_3^{2-}
Monohydrogen phosphate ion	HPO_4^{2-}	4.47×10^{-13}	Phosphate ion	PO_4^{3-}
Water	H_2O	1.81×10^{-16}	Hydroxide ion	OH^-
Conjugate acid	**Formula**	**K_a of acid**	**Base**	**Formula**

Increasing acid strength → Increasing base strength

REFERENCES

[1]   Metz, C.; Castellion, M. E. *Chemistry: Inorganic Qualitative Analysis in the Laboratory*; Academic Press: Now york, **1980**, p. 130.

[2]   Baum, S.J.; Scaife, C.W.J. *Chemistry, a Life Science Approach,* 2nd ed; Macmillan: New York, **1980**, p. 828.

[3]   Beran, J.A. *Laboratory Manual for Principles of General Chemistry,* 10th ed; Wiley: Hoboken, NJ, **2014**, p. 448.

[4]   Mils, J.L.; Hampton, M.D. *Microscale and Macroscale Experiments for General Chemistry*; McGraw-Hill College, **1991**, p. 307.

[5]   Vogel, A.I.; Svehla, G. *Vogels Textbook of Macro and Semimicro Qualitative Inorganic Analysis,* 5th ed; Longman: London, **1979**, p. 605.

[6]   Chang, R.; Goldsby, K.A. *Chemistry,* 10th ed; McGraw-Hill: London, **2013**, p. 1085.

[7]   Ebbing, D.D.; Gammon, S.D. *General Chemistry,* 9th ed; Houghton Mifflin: Boston, Mass, **2009**, p. 1030.

[8]   McQuarrie, D.A.; Gallogly, E.B.; Rock, P.A.; Mcquarrie, C.H. *General Chemistry,* 4th; University Science Books: Ram, **2011**.

[9]   Brescia, F.; Arents, J.; Meislich, H.; Turk, A. *Fundamentals of Chemistry, Laboratory Studies,* 3rd ed; Academic Press: New York, **1975**, p. 306.

SUBJECT INDEX

www.ingramcontent.com/pod-product-compliance
Lightning Source LLC
Chambersburg PA
CBHW041450210326
41599CB00004B/199